T0303888

RUSSIAN TRANSLATIONS SERIES

(continued)

ROCK BREAKAGE
BY BLASTING

ROCK BREAKAGE BY BLASTING

M.I. Petrosyan

RUSSIAN TRANSLATIONS SERIES
105

A.A. BALKEMA/ROTTERDAM/BROOKFIELD/1994

Translation of : *Razruzhenie gornikh porod pri vzrivnoi otboike*,
Nedra, Moscow, 1991

Translator   : Mr. S. Sridhar

Technical Editor : Prof. A.K. Ghose

General Editor   : Ms. Margaret Majithia

ISBN 90 6191 902 9

Distributed in USA and Canada by : A.A. Balkema Publishers, Old Post
Road, Brookfield, VT 05036, USA.

Preface

The solution to many pressing problems of technical progress is directly linked with increasing the production of ferrous, non-ferrous, precious and rare metals, much of which is exploited from underground mines under complex geomining conditions. Attempts to upgrade the quality of ore and fuller extraction have been sought through development of effective mining methods using cemented backfill [13].

At present, perfecting the technology of ore exploitation is mainly directed towards widening the field of application of the method of mining with backfill and trackless equipment [4]. The stipulations imposed on fragmentation by blasting have become concomitantly more stringent. The task of blasting lies not only in separating the ore from the massif and its subsequent breakage to a size that maximises the output of free-steered vehicles, but also in increasing the yield of average fractions favourable for the very widely used method of autogenous grinding. In this respect, additional requirements are imposed on the stability of the massif around the face represented by host rocks or backfill.

A wide spectrum of research and development work in the mining of non-ferrous metals is reflected in the publications of M.I. Agoshkov, D.M. Bronnikov, G.P. Demiduyk, V.R. Imenitov, M.V. Kurleni, N.F. Zamesov, E.G. Baranov, V.N. Mosints, V.N. Skubiy, S.V. Kuznetsov, N.G. Dubynin, M.N. Tsygalov and others.

The adoption of fragmentation by blasting in methods of working with backfill has stimulated research on the control of blast effects. Since the problem of enhancing the quality of ore breakage from the massif and changing its granulometric composition cannot be solved directly under field conditions, various laboratory modelling methods have been developed [30, 31].

The method of equivalent materials is the most effective among the various modelling techniques. This method helps in establishing the field of rational blasting parameters and in evaluating blasting efficiency based on the end result—granulometric composition of ore and stability of the massif around the face. The massif outside the line of perimeter holes is termed the massif around the face.

The main task of blasting mechanics is to study the breakage process of rocks. Blasting processes in rocks and soils have been theoretically analysed by L.I. Sedov, G.P. Cherepanov, O.E. Vlasov, G.I. Pokrovskii, G.M. Lyakhov, E.I. Shemyakin, S.S. Grigoryan, V.N. Rodionov, V.M. Kuznetsov, V.M. Komir, N.Ya. Repin, V.V. Adushkin, V.S. Nikiforovskii, E.N. Sher, A.L. Isakov, V.N. Nikolaevskii, I.A. Sizov, V.M. Tsvetkov, A.A. Spivak, Brod, Taylor, Brookweigh, Allen, Passman, Davison, Curran and others

Despite the availability of various publications, the physical concepts of the mechanism of breakage are divergent. Attention needs to be paid to the factors which impede further research into the process dynamics of crack formation and crushing of rocks during blasting. Complications arise with regard to the solution of methodological problems. It is practically difficult to record the growth of cracks in optically inactive media and particularly in rocks and concrete. Attempts to simplify the solution to this problem gave rise to modelling of the rock breakage process using optically active materials (epoxy, polymers etc.).

The traditional approach to assessing the blast-induced mechanical effect is based on the action of elastic wave on the medium being studied. The wave elastic front can be observed and recorded in rocks only at pressures of the order 10^4 MPa. According to V.N. Rodionov, the front of the disturbance happens to be the elastic wave; the dimensions of the front area are very small and in a blast of explosive charges this area is completely absent. Given such conditions, any attempt to consider the phenomenology of rock breakage based on loss of transparency in material, led necessarily to improper physical representations. In reality, the loss of material transparency is affected by breakage in microscopic volumes and also by the action of stress wave. Here two noteworthy assumptions are made. The first is that the delay in breakage, governed by preparation of the medium for crushing, is not completely taken into account. The second is the inadequate resolution capability of the recording devices. In certain publications, for example [7], attempts have been made to assess the development of the breakage process of the entire rock massif in the active phase of stress wave, based on the formation of macrocracks at the drillhole contour. According to other publications, the formation of macrocracks by the action of shock waves further propagates into the wave phase of the blast. This has not been substantiated. Research work conducted by Isakov and Sher [12] showed that local fracturing at the hole contour occurs due to the effect of the blast wave, while during the phase of piston effect of gases, radial cracks develop in directions determined by the sites of these local fractures. During this phase the ultimate rate of crack movement is equal to 650 m/s. Rodionov [28] observed that the said wave cannot be the characteristic feature of the initial event of the blasting effect on the medium, i.e., it cannot be, the primary cause for the breakages that occur.

According to investigations conducted by Curran and colleagues [39], at a specific distance from the source of disturbance at which stresses significantly weaken, the rock may experience different stages of crack initiation, crack growth and crack coalescence but it remains at the same place in the form of a damaged massif of lower strength. If the plasticity state of the rock massif is considered as the prefailure state [15], it then becomes obvious how important is a knowledge of the phenomenological criteria for transition of a crack into its rapid growth phase.

Instability of cracks is usually determined by two types of criteria—criterion of local stress and energy criterion. As an illustration, Griffith's criterion may be used, which is based on the first law of thermodynamics. The following basic concepts are inherent in Griffith's theory. Defects in the form of microcracks are present in rocks and when subjected to loading, stress concentration is caused inducing propagation of cracks; this process finally results in the failure of the macroscopic volume. The interrelationship between the crack length, surface energy and applied stress is given by

$$\sigma_0 = \sqrt{\frac{2E\gamma_0}{\pi(1 - \mu^2)l}}$$

where E—elasticity modulus;

   μ—Poisson's ratio;

   γ_0—surface energy of a freshly exposed crack;

   l—crack length.

The crack loses its equilibrium at stresses exceeding σ_0, which results in dynamic growth.

G. Irwin and E. Orovan developed the concept of quasi-brittle failure in which, for the first time, the localised zone of plastic relaxation was taken into account. The existence of this zone is associated with a certain amount of energy consumption.

G.I. Barenblatt developed the theory of normal rupture of cracks. He postulated the following hypotheses: the dimension of the crack tip is very small compared to the dimension of the entire crack; distribution of displacements at the tip does not depend on operating loads and for a given material under specific conditions it is always constant; opposite surfaces of the crack close smoothly at its contour and tensile stress at the contour of the crack is finite. S.A. Christianovich proposed the condition of finiteness of stresses: at the crack tip the stress in a rock must be finite, otherwise the crack could not have closed. This condition is taken as the basis for determining the dependence of crack length on various forces.

G.P. Cherepanov observed that the aforesaid investigations on the theory of cracks were devoted to cracks of normal rupture. This is associated with the situation that breakage under tensile stresses is often encountered in practice. It has been confirmed that a crack formed under conditions of compression in a rock massif, always happens to be a crack of transverse shear. For example, in mines at the surface of supports or pillars, it has an arbitrary shape, starting from a straight line and ending with a step or curve of the 'ladder' or 'herringbone' type. Within the framework of this scheme, the failure criterion for a brittle body is given by

$$\sigma_f = 2\sqrt{2}G_0/(\pi\sqrt{l}),$$

where G_0—shear modulus;
l—crack length.

When the load attains the ultimate value of σ_f, the equilibrium of the body destabilises and any subsequent increase in load leads to dynamic growth of the cracks.

The aforesaid criterion postulates the onset of failure at a certain ultimate critical situation in time. Fadeenko [34] confirmed that the time-related criterion of failure has long been known; however, it did not find wide acceptance due to the relative limitations of range of deformation rates adopted in technological processes.

The foregoing example shows that the concept of coalescence of microcracks is the basis for the initiation of failure [15, 16, 19]. Such deformation criteria of strength are considered one of the principal characteristics of the process—the loading time and subsequent failure. It thus follows that solid bodies fail at a specific level of stress, when deformations reach critical values [15].

Researchers have been paying attention in recent years to the problem of intensified ore preparation. A key role is played by the loosening of ore by blasting. Based on the coalescence concept, Repin [8] showed the possibility of directing the effect of a blast on the properties of ores and mineral components. As a result, the quality of concentrate is upgraded, recovery of useful components from ores is increased and losses during beneficiation are reduced.

One of the salient features of the technology of rock breakage by blasting is the attainment of requisite quality of fragmentation, taking into account the stability of the massif around the face. The breakage process of the massif around the face is determined mainly by the structure of enclosing rocks or stowed material and is characterised by the development of micro- and macrocracks due to the effect of tensile pulse loading governed by the reflection of the stress wave from the exposed face. The slabbing mechanism of failure was studied in detail by Nikiforovskii and

Shemyakin [15]. Research in this direction has presently intensified. More efforts are needed to create new and effective methods of rock breaking by blasting.

Research work on the dynamics of crack growth is mainly devoted to problems of stability of directed cracks and their limiting velocity. Experimental data on the types of branching of cracks and rate of enlargement of the surface of cracks (rate of movement) due to blasting are not available in published literature. Consequently, it is not possible to fully assess the mechanical effects induced by blasting explosive charges. An assessment of blast mechanics according to the end-results as well as dynamics of formation process, growth and branching of cracks are essential for further development of the theory and practice of breakage by blasting. Precisely such an approach distinguishes the present book from earlier ones. Solving a series of methodological problems arose in connection with the conduction of complex investigations into the process of breakage by blasting.

Intensive study in recent years of the kinetics of rock breakage by blasting has led to the development of blasting mechanics, i.e., dynamics of processes of crack formation and fragmentation [3, 15, 16, 21, 25, 36, 37]. The established types of branching in cracks indicate that some relationship exists between the rate of tearing apart of the surfaces of cracks and the development of the failure process itself.

The influence of deformation on the amplitude of blast-induced stresses in the massif around the face in tight conditions is of particular interest. Research in this direction was conducted earlier by Jones, Lermit, Roche and Berg. The experimental results of blast loading showed that the stress, corresponding to the limit of violating the continuity in material, is the most important feature of the medium under study. In this monograph, special attention is paid to the fractional participation of the blast wave and piston effect of gases in a quantitative ratio, which is evaluated by their contribution to failure expressed in micro- and macroscopic volumes.

If multirow short-delay blasting is adopted, taking into account the real rate of failure of the massif, then new possibilities arise for rock fragmentation and maintenance of the stability of the massif around the face contour. Upon considering the rate of crack movement and velocity of widening of its surfaces (motion of massif), in particular, it becomes possible to effectively use the method of ore breakage by blasting of drill holes with presplitting of the narrow stoped-out area.

In conditions of intensive broadening of the field of application of the methods of working with backfill and usage of free-steered vehicles, the study of parameters of breakage by blasting in models can give a remarkable effect.

Verification of the results of modelling in field conditions is an important phase in any investigation. The results of laboratory and field experiments enable evaluation of the error in modelling of breakage by blasting. A criterion for assessing the work done by a blast has been presented in this book, such as specific consumption of (explosive) gases for breaking ore (rocks). Using this criterion, the results of ore (rock) fragmentation in the model and under actual conditions can be compared. Formulae are suggested herein for accurately determining delay intervals. These may find widespread usage in blasting practice.

T.G. Gasparyan and V.V. Polusyan collaborated with the author in experimental investigations. Their collaboration made it possible to conduct a large number of experiments.

The author will gratefully accept any suggestions or remarks regarding the contents of this book.

Contents

1

Technique of Modelling for Rock Breakage by Blasting

1.1 Tasks and Methodology of Investigations

Numerous experimental studies in recent years have been directed principally towards improving the operations of backfilling [9, 18, 20, 31, 38]. Some of the serious impediments encountered in the development and perfection of the system of working and upgrading the effectiveness of ore preparation are drilling holes of small diameter in the massif and breakage of fragments of specific size loosened in the process of blasting. In spite of such difficulties, drilling and blasting constitute the principal method for breaking hard minerals [5]. The quality of broken rock and productivity depend mainly on the extent of controlling the blast effects.

It is necessary to consider the following special features of the rock breakage process by blasting, which impose certain requirements on limiting the field of utilisation of its parameters. Firstly, the deposits are heterogeneous vis-à-vis the physicomechanical properties of ore and rocks with a different degree of disturbance in the massifs, which significantly affects the breakage processes. Secondly, the blasts affect the stability of the backfilled material, particularly when it is exposed. These factors complicate the choice of rational values for blasting parameters in underground metal mining practice.

In this connection, a major task of investigation involves the development of the physical fundamentals of rocks breakage by blasting, the methodology of reproducing specific complicated situations in models, studying them in models and developing methods of computing optimal parameters of the ore-breakage process, taking into consideration the dynamics of brittle fracture.

Complex methods of investigation have been adopted to solve such problems. They comprise the study of kinetics of rock breakage by blasting, laboratory modelling of blasting with invariant stresses, field blasts to verify the results of laboratory modelling and establishment of rational parameters of ore breakage by blasting.

Study of kinetics of rock breakage by blasting is undertaken to solve the following problems:

1. Determination of the rate of movement of a crack and rate of opening of its surfaces.
2. Assessing crack formation in blasts of charges of different designs.
3. Establishing fractional participation of the blast wave and piston effect of the detonation products in breaking a massif.

The kinetics of rock breakage was studied by the author by the method of recording cracks, including those induced by blasting, with synchronous high-speed photography using the time loop LV-04 equipped with an electro-optical transducer (image converter tube) (EOP) of the type UMI-92 or UMI-93 [22]. The rate and direction of movement of a crack were controlled by breaking graphite rods (repers) installed in the model block.

Based on an analysis of the photographic frames, the extent of growth of cracks over time, types of branching etc. were established.

Laboratory modelling of rock breakage by blasting is done by the method of equivalent materials with the objective of solving the following key problems:

1. Establishing the laws of breakage of the ore massif by blasting.
2. Determining the relationship between breakage parameters and stresses developed in adjacent rocks when a blast wave (pulse) passes through and its effect around the face contour.

To solve the defined problems, the material for the model is selected after studies on the strength and elastic characteristics of the ore and rock of an experimental section in a particular mine have been done. While selecting the equivalent material, the well-known relationships between the strength and elastic properties of materials in the model and those of the prototype are used, which are necessary and sufficient for fulfilling conditions of similarity.

Since modelling is undertaken on a reduced scale, the scaling of all geometrical and physical parameters from the prototype (actual) to the model (and conversely) is effected considering the similarity conditions and maintaining the corresponding dimensions. The methodology of modelling suggested here helps in obtaining comparable results of rock breakage by blasting in a model as well as in real conditions.

While conducting experiments to study the laws of ore breakage, the blast pulse at different points in the model is recorded simultaneously with the help of sensors. Both compressive and tensile stresses are recorded. This is the main distinguishing feature in the method suggested. Based on an analysis of oscillograms of the blast pulse, the stresses developed in the layer being blasted and in enclosing rocks are quantitatively assessed.

Field experiments are essential to verify the results of modelling and to establish rational values for ore-breakage parameters. Considerable deviation is observed between the granulometric composition of broken ore in the prototype and in the model. This discrepancy is basically associated with the difference in rigidity of materials of the model and the prototype and also the discontinuities in the rock massif. The discrepancy is further attributable to differences in properties of the explosives used in the model and in the field. It is necessary to develop a criterion for the work done by a blast—in the form of specific consumption of explosive gases used in breakage—in order to quantitatively evaluate the discrepancy. This criterion enhances the reliability of comparative evaluation of the results of ore breakage in the model and in the prototype.

Field trials are also essential for solving a series of other problems, in particular those related to determination of the effect of quality of fragmentation on productivity, totality of extraction etc.

Technoeconomic evaluation is the final stage in the investigation. Results of experiments and time-motion study data are the inputs for such an evaluation.

While assessing the parameters of rock breakage by blasting, the quality of fragmentation is evaluated by the yield of boulders, medium-size fragments and ore fines. If boulders are excessive, the output of loading-hauling machines decreases. Given the requirements of autogenous grinding in crushers, fragments below 50 mm in size should not exceed 25%.

After evaluating the rational schemes of breakage, that one amongst them which incurs the least expenditure in mining one ton of ore is selected.

1.2 Principal Equations of Correlation and Criteria of Similarity

Authenticity of laboratory studies is achieved by conforming to the principles of modelling, namely, similarity of prototype and model phenomena. N.F. Zamesov studied the principal laws of ore breakage by means of models taking into account the following:

—Geometrical similarity of major elements of layer (bench) being blasted and disposition of charges;

—Energy-wise similarity of model and prototype;

—Similarity of granulometric compositions of the broken ore in the model and the prototype.

Geometrical similarity is achieved by reducing the dimensions of the layer being modelled by n times (scale of modelling $n < 1$), i.e.,

$$l_m/l_a = n, \quad b_m/b_a = n, \quad W_m/W_a = n, \quad S_m/S_a = n^2, \quad V_m/V_a = n^3.$$

$$(1.1)$$

where *l*—characteristic dimension (length of layer);
    b—spacing of charges;
    W—line of least resistance (LLR);
    S—surface of layer;
    V—volume of layer.
The subscript 'm' refers to the model while 'a' refers to the actual (prototype).

Considerable difficulties are encountered, however, in reproducing an equivalent blast pulse so as to maintain conformity between the blast energy and granulometric composition of the broken ore in both the model and the prototype. To solve such a problem N.F. Zamesov recommended the criterion:

$$U_m/(R_m S_m r_m) = U_a/(R_a S_a r_a) = \text{const};$$

which is based on Rittinger's law.
Here, *U*—blast energy;
    r—characteristic linear dimension of an elementary volume;
    R—resistance of the medium against deformation.
While modelling, the structure of the material does not depend on global dimensions (scale of modelling). In such conditions, the hypothesis of 'almost-near effect' is justified, as propounded by G.P. Cherepanov. It is based on the physical fact that forces of interaction of elementary particles decrease very fast with an increase in distance between them. At distances *r* they can be neglected.

Zamesov's criterion shows that in breaking material in the model equal in strength to the prototype, it is not possible to obtain granulometric compositions similar to those in actual conditions as the specific energy ΔU in the model needs to be enhanced the same number of times by which the dimension of the model was reduced from the actual. Moreover, it is not possible to maintain correspondence between strength and stresses in the model material, due to the effect of gravitational force, when the size of the block is changed. For modelling rock breakage by blasting, it is therefore necessary to use material in the model whose elasticity modulus and ultimate strength closely approximate the scaled reduction of the model from the actual.

The massif is characterised in its initial state by the elasticity modulus E, Poisson's ratio μ, the medium's resistance to deformation R and density γ. The system of decisive parameters is thus: μ, E, R, γ, l. Of these five, two have no dimensions and therefore the three dimensionless ratios—μ, $R/\gamma l$ and $E/\gamma l$—will serve as the basis for mechanically similar states of the layer.

The numerical values of the dimensionless ratios in these two phenomena are equal.

$$\mu_m = \mu_a = const;$$
$$R_m/(\gamma_m l_m) = R_a/(\gamma_a l_a) = const;$$
$$E_m/(\gamma_m l_m) = E_a/(\gamma_a l_a) = const. \qquad (1.2)$$

The similarity criteria require that these ratios be equal both in the model and the prototype.

If we assume that $R_m \neq R_a$ and $E_m \neq E_a$, i.e., switch to the method of equivalent materials, we obtain

$$R_m/R_a = \gamma_m l_m/(\gamma_a l_a) \text{ and}$$
$$E_m/E_a = \gamma_m l_m/(\gamma_a l_a). \qquad (1.3)$$

Correlation between the strength and elasticity constants of materials in the model and in the prototype, at $\gamma_m = \gamma_a$, is of the type:

$$R_m = R_a n; \qquad (1.4a)$$
$$E_m = E_a n; \qquad (1.4b)$$
$$\mu_m = \mu_a. \qquad (1.4c)$$

In this case

$$U_m/U_a = R_m S_m/(R_a S_a). \qquad (1.5)$$

Solving equations (1.4a) and (1.5), we get

$$U_m = U_a n^3. \qquad (1.6)$$

From the ratio $U_m/V_m = U_a n^3/(V_a n^3)$ at $\gamma_m = \gamma_a$, we obtain

$$\Delta U_m = \Delta U_a. \qquad (1.7)$$

Thus, in the modelling of rock breakage by blasting using the method of equivalent materials the necessary condition for obtaining quality of fragmentation close to that of the actual, is the equality of specific energy of explosive both in the model and the prototype.

If commercial explosives (ammonite, grammonite etc.) are used in actual conditions and TEN is used in the model (with a charge density of 1 g/cm^3), then charges are recalculated according to the relation

$$d_m = d_a n, \qquad (1.8)$$

where d_m and d_a—diameter of charges used in the model and the prototype.

However, such an approach leaves open the question of the effect of blast around the face contour. Hence the need arose for taking into account an additional similarity criterion based on wave propagation processes.

A study of the nature of the breakage process of enclosing rocks (walls of winning rooms) showed that breakage by blasting is accompanied by crack formation in the interior of the massif and breakage of enclosing rocks or filled material. The perimeter charges at the contact or near it act mainly on the process of crack formation. Let us assume that at a given point the rock element behaves as an elastic body. Then, based on the known postulate of the theory of elasticity [32], it may be assumed that the maximum stress σ_{max} which will develop in the massif when the blast wave passes through and the maximum velocity of particles v_{max} are related to each other by

$$\sigma_{max} = \gamma C_p v_{max}, \tag{1.9}$$

where γ—density; C_p—velocity of longitudinal wave.

Considering (1.9), the following relationship is justified

$$\sigma_{max_m}/\sigma_{max_a} = \gamma_m C_{p_m} v_{max_m}/(\gamma_a C_{p_a} v_{max_a}). \tag{1.10}$$

Let us consider the similarity criteria of states of the massif around the face by the given relationship (1.10).

From the theory of elasticity, we know that

$$C_{p_m} = \sqrt{\frac{E_m(1 - \mu_m)}{\gamma_m(1 + \mu_m)(1 - 2\mu_m)}}; \tag{1.11}$$

$$C_{p_a} = \sqrt{\frac{E_a(1 - \mu_a)}{\gamma_a(1 + \mu_a)(1 - 2\mu_a)}}. \tag{1.12}$$

where E_m and E_a—elasticity moduli of materials of the model and the prototype;

γ_m and γ_a—density of materials of the model and the prototype.

Taking into account (1.4b), at $\gamma_m = \gamma_a$ and $\mu_m = \mu_a$, and by solving (1.11) and (1.12), we obtain an important criterion that determines the properties of equivalent materials,

$$C_{p_m} = C_{p_a} \sqrt{n}. \tag{1.13}$$

Considering geometrical similarity, let us establish the relationship between maximum particle velocities in the model and the prototype, i.e.,

$$v_{max_m} = v_{max_a} \sqrt{n}. \tag{1.14}$$

Taking into account (1.13) and (1.14), at $\gamma_m = \gamma_a$, we obtain at similar points in the model and the prototype

$$\sigma_{max_m} = \sigma_{max_a} n. \tag{1.15}$$

The relationship (1.15) is valid for modelling the process of breakage by blasting using equivalent materials. The idea of using this method originated from V.R. Imenitov. Subsequently, he developed a set of methodological postulates of modelling the dynamics of caving and ore extraction.

Despite the refinements achieved in models made from equivalent materials, errors invariably creep in due to the differences in properties of explosives used in the model and the prototype as well as the rigidity of both, which depends on the elastic properties of rocks, their structural features and mineral composition. Due to such conditions, the need arose for modelling breakage by blasting separately for each mine (district, block). It should be noted that modelling the structure of a massif is complicated and a problem not readily solved.

According to investigations by L.I. Baron, a relationship exists between strength of rock (ore) and specific consumption of explosive q,

$$q_1 = q_2 R_1 / R_2, \qquad (1.16)$$

where q_1—specific consumption of explosives for breaking rocks with resistance R_1 against deformation;
q_2—specific consumption of explosives for breaking rocks with resistance R_2 against deformation.

However, the field of application of (1.16) is limited by the properties of explosives. If the ore is blasted by explosives of different properties, the relationship is violated and criterion (1.16) cannot be used. Experience has shown that in such conditions it is advisable to evaluate blast efficiency by using the criterion V_g—specific volume of gases released during the blast, breaking ore, l/T (l/m^3). The criterion V_g is necessary for putting in order the experimental data, in other words to reduce the scatter of fragments. The specific consumption of gases for breaking rocks is the characteristic product $q V_{g_*}$, (q—specific consumption of explosives for rock breakage, V_{g_*}—quantity of gases (l) produced by blasting 1 kg explosive). Therefore, it is handy to alter the relationship (1.16). Substituting V_g in the place of q, we get

$$V_{g1} = V_{g2} R_1 / R_2. \qquad (1.17)$$

From (1.17) it follows that $V_{gm} = V_{ga} n$, which is not possible. Thus, the need for adopting the method of equivalent materials for modelling is confirmed.

Expressing V_g by the volume of gases V_1 released during the blasting of charges and using geometrical similarity, we get

$$V_{m_1} = V_{a_1} n^3. \qquad (1.18)$$

Correspondingly, knowing the relation between volume of materials broken in the model and the prototype (1.1), at $\gamma_m = \gamma_a$, and dividing (1.18) by (1.1), we get

$$V_{g_m} = V_{g_a}. \qquad (1.19)$$

The equality of specific volume of gases released in the model and the prototype happens to be the necessary condition for comparative evaluation of the granulometric composition of the ore blasted by explosives of different properties.

The delay interval is a key parameter that decides the effectiveness of breakage by blasting. Let us establish the correlation between delay intervals in the model and the prototype. The rate and time of propagation of longitudinal wave and LLR are related by

$$C_{p_m} = \frac{W_m}{t_m} \text{ and } C_{p_a} = \frac{W_a}{t_a},$$

where C_{p_m} and C_{p_a}—velocity of propagation of longitudinal wave in the model and in the prototype;

W_m and W_a—LLR in the model and in the prototype;

t_m and t_a—time of propagation of longitudinal wave in the model and in the prototype.

Taking this into account in (1.13), we get

$$t_m = t_a \sqrt{n}. \qquad (1.20)$$

The recalculation of delay intervals from the model to the prototype is undertaken according to (1.20).

Similarity of ore fragmentation and blast effect around the face is ensured by selection of equivalent materials, taking into account the additional criterion (1.13).

The rational delay interval between rows of charges and also between charges themselves in a row is determined based on the rate of movement of crack v_T and rate of opening of its surfaces v_b. The values of v_T and v_b for a slot are established experimentally.

It should be borne in mind that the results of investigations of slot parameters in the model and the dynamics of crack formation (fragmentation) largely depend on the physicomechanical properties of the medium. In this case it is necessary to achieve similarity between elastic properties of materials in the model and the prototype; otherwise, the slot formation by blasting is not possible. Consequently, special attention is paid in this monograph developing suitable equivalent materials.

1.3 Selection of Equivalent Materials

Equivalent material for modelling rock breakage by blasting should be selected after taking into account the similarity of strength and elastic properties of the rocks being modelled, as determined by the correlations (1.4a), (1.4b), (1.4c) and (1.13). Depending on the strength of rocks and the scale of modelling, the required strength of equivalent material is assigned. No special difficulties are encountered in achieving similarity in strength properties of these materials; however, the solution to correlation (1.13) that determines the similarity of elastic properties of materials of the model and the prototype is somewhat complicated.

Hence an equivalent material of another composition was chosen as filler, which is characterised by a high acoustic absorption coefficient, in order to preserve the similarity of elastic properties. A microseismic wave is generated in the process of formation of microcracks. This model corresponds more completely to the prototype as the rock massif contains microcracks in its initial state. The need for studying the physicomechanical properties of ores and enclosing rocks in the deposits, for which modelling is being done is readily apparent.

Acoustic rigidity is the principal measure of resistance of rocks to breakage (crushing) under dynamic loads. It is determined by the expression $A = \gamma C_p$. If γ_m is assumed to be equal to γ_a, then the similarity condition amounts to retaining equation (1.13).

Generally, the rate of propagation of longitudinal wave C_p can be expressed as a function of certain parameters, which determine the physicomechanical properties of materials:

$$C_p = \xi(E, \gamma, \mu).$$

Here E—dynamic modulus of elasticity;
γ—density;
μ—Poisson's ratio.

The velocity of the longitudinal wave is given by

$$C_p = \sqrt{\frac{E(1-\mu)}{\gamma(1+\mu)(1-2\mu)}}, \qquad (1.21)$$

while the dynamic modulus of elasticity is expressed as

$$E = \frac{\gamma(1+\mu)(1-2\mu)C_p^2}{(1-\mu)} \qquad (1.22)$$

The dynamic elasticity modulus can be graphically determined with a tangent at the origin of the static stress-strain curve under a high rate of loading. The dynamic method in which loads of very small amplitude act for a very short period helps determination of the constant value of elasticity modulus of material that is stress independent. Moreover, in such experiments internal ruptures do not occur in the material and consequently no structural changes occur. For this reason, the dynamic modulus of

10

elasticity helps one to judge more appropriately whether deformations are taking place in the material rather than in the static modulus.

The velocity of longitudinal wave is determined by means of a portable pulse-measuring apparatus IPA-4 through direct measurement of the propagation velocity of ultrasonic pulse through a cube having an edge of 7 cm or a prism of 7 × 7 × 14 cm made from rock or equivalent material.

The relationship between acoustic rigidity and coefficient of hardness of rocks is determined by the equation (Fig. 1)

$$A = 2.3 \times 10^5 f^{0.58} \qquad (1.23)$$

with a correlation coefficient of 0.9. Results of investigations of the relationship between A and f indicate the existence of a close link between them, which enables one to use them for solving practical problems.

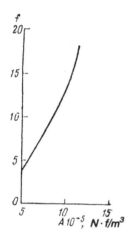

Fig. 1. Relationship between coefficient of hardness of rocks and acoustic rigidity

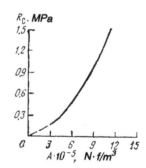

Fig. 2. Relationship between static strength of rocks in the experimental section and their acoustic rigidity

Values of Coefficient of Hardness of Rock, f, according to M.M. Protod'yakonov Scale and Acoustic Rigidity of Rocks, A

f ...	2.5–3.5	3.5–4.5	4.5–5.5	5.5–7
$A \times 10^{-5}$, N·f/m³	3.3–5.4 (5.21)	6–6.88 (6.46)	5.5–5.18 (6.48)	6–9.23 (8.2)
f ...	7–9	9–12.5	12–17.5	
$A \times 10^{-5}$, N·f/m³	5.9–9.51 (8.75)	7.5–11.15 (9.2)	8.2–11.85 (9.72)	

Note: Average values are given in parentheses.

In order to model the process of rock breakage by blasting using equivalent materials, the physicomechanical properties of rocks in a specific mine were studied (Table 1). Ore and enclosing rocks did not differ significantly in their physicomechanical properties; the contours of mineable reserves in the zone were established by sampling from mine workings. Therefore, identical equivalent material for the ore and enclosing rocks was used.

It was experimentally established that the properties of the host rocks being modelled and the ores investigated from other similar deposits are characterised by the following indices:

σ_c = 80 to 120 MPa; γ = 2.5 to 2.6 g/cm³;
μ = 0.19 to 0.24; E = (2 to 6) ×10⁴ MPa:
C_p = 2760 to 4910 m/s; A = (4 to 16.2) ×10⁵ N·f/m³.

The relationship between strength and acoustic rigidity of the actual material was determined by an equation of the type:

$$R_c = 0.218 \times 10^{-5} A^{1\ 73} \qquad (1.24)$$

with a correlation coefficient of 0.9.

During the investigation of acoustic properties of non-metallic materials, the idea of creating a new composition of equivalent material was conceived [2]. The coefficient of absorption of sound wave in concrete is known to be equal to 0.015. The wave absorption coefficients of materials used as filler after crushing (milling) are always greater than that of

Table 1

Rocks	γ, g/cm³	$R \times 10^{-2}$, MPa	C_p, m/s	μ	$E \times 10^{-5}$, MPa	$A \times 10^{-5}$, N f/m³
Carbonaceous talc	2 65	0.24	2595	0.13	0.175	4.45
Carbonaceous talc with carbonates predominant	2.4	0 37	4100	0.24	0 34	8 1
Anorthosite, carbonaceous talc with talc predominant	2.65	0 56	3450	0 20	0.25	7.45
Gabbro (unaltered), slightly carbonaceous and chlorinated gabbro	2 7	0.87	3840	0 17	0 37	10 4
Unaltered gabbro, carbonate-serpentine rocks, slightly chlorinated gabbro	2 5	1.11	4190	0.19	0.48	10
Hornblende gabbro, rich quartzitic foliaged rocks	2.8	1 41	4430	0 20	0.47	13 3
Rich quartzitic foliaged rocks	2 9	2 3	4110	—	—	10.4

concrete. Hence compositions based on powdered marble have a high coefficient of wave absorption, since this index for marble is equal to 0.01.

The composition and properties of equivalent materials are given in Tables 2 and 3. Composition III has the lowest elastic properties. The velocity of longitudinal wave in the material based on powdered white marble is considerably higher (1090 m/s); an analysis of its substance showed that this material was metamorphosed limestone. In the equivalent material based on quartz sand (composition I), the highest velocity of longitudinal wave observed was 1620 m/s.

The properties of equivalent material based on powdered black marble are characterised by the following indices:

$$\gamma \quad = 2.5\text{--}2.6 \text{ g/cm}^3; \qquad \sigma_c/\sigma_p = 9\text{--}13;$$
$$E = (0.8 \text{ to } 1.0) \ 10^3 \text{ MPa}; \qquad \mu = 0.19\text{--}0.22;$$
$$C_p \qquad = 600\text{--}650 \text{ m/s}; \qquad A = (1.5 \text{ to } 1.7) \ 10^3 \text{ N·f/m}^3.$$

Substituting the values of R, E, μ and C_p in correlations (1.4a), (1.4b), (1.4c) and (1.13), we found that the properties of the equivalent material (composition III) chosen for modelling rock breakage by blasting, are well within the required limits vis-à-vis properties of the ores and rocks under study.

Sometimes, it becomes necessary to model the process of breakage by blasting ores having a density of 3–4 g/cm^3. For this purpose,.mixes

Table 2

Component	Density γ, g/cm^3	Composition, % by weight		
		I	II	III
Quartz sand	1 59	51	—	—
Powdered white marble	1.75	—	60	—
Powdered black marble	1.80	—	—	70
Cast iron filings	3.03	35	25	22
Cement M-300	1.56	14	15	8
Water in the mass	1.0	10	10	10

Table 3

Parameter	Composition		
	I	II	III
Density, g/cm^3	2.5–2.6	2.5–2.6	2.5–2.6
Ultimate compressive strength, MPa	1 7–1.8	1 7–1.8	1.7–1.8
Ultimate tensile strength, MPa	0.24	0.2	0.14–0.18
Elasticity modulus, MPa	6.5×10^3	2.5×10^3	$(0.8 \text{ to } 1.0) \times 10^3$
Poisson's ratio	0 07	0 19	0.19–0.22
Velocity of longitudinal wave, m/s	1620	1090	600–650
Acoustic rigidity, N·f/m^3	4×10^5	2.7×10^5	1.6×10^5

of heavier equivalent materials based on black marble of 2.7–4 g/cm^3 density were developed, which helps to widen the field of application of the method of equivalent materials.

Composition of Heavier Equivalent Materials (%)

Powdered marble	62	47	31	23
Cast iron filings	30	45	61	69
Cement	8	8	8	8
Water in hard components	9	8	7	6
Density of composition, g/cm^3	2.7–2.8	3–3.1	3.5–3.6	39.4*

*Sic; so given in Russian original—General Editor.

The velocity of the longitudinal wave in the material increases (800–2080 m/s) when the quantity of iron filings is increased. This should necessarily be borne in mind while modelling rock breakage by blasting.

1.4 Methods for Recording Stresses Induced by a Blast

Various methods for measuring stresses induced by blast in a hard medium have been developed in the last decade but deviations in the results of the recorded parameters were also detected. Reliable results are essential to selecting an appropriate method of recording the blast-induced stresses in the model and the prototype [1]. The blast-induced stresses in marble (5), granite (3, 4, 6, 7), rosin (2), gypsum sulphite (1), concrete (8) which were recorded in [1] are shown in Fig. 3. For the ore and the same rock (granite), stresses of different magnitudes were recorded at the same relative distances r/r_0 (r—distance from the charge centre to the stress recording site, r_0—radius of charge), when different explosives were used. Maximum stress σ_{max} was equal to 34.5 MPa for TEN [$C_5H_8(ONO_2)_4$] at $r/r_0 = 30$ (6), while for ammonite 6ZhV it was 12.5 MPa (4). But in a blast of ammonite 6ZhV in granite (7) at a distance of $r/r_0 = 90$, σ_{max} was likewise found to be 12.5 MPa. When a charge of TEN was blasted in marble (5) and granite (6) at a similar relative distance $r/r_0 = 15$, σ_{max} was recorded as 119 and 200 MPa respectively. Experimental curves 3, 4, 5, 6, 7 were drawn based on the data of A.N. Khanukaeva. In laboratory experiments (curves 5, 6), stresses were recorded on a peizoelectric transducer based on seignette salt; for a blast in a drill hole (curves 4, 7) and concentrated charges (curve 3) in an open-cast mine, the stresses were recorded on vibrographs. The deviation in magnitude of recorded stresses was due to the properties of the medium and conditions of recording.

It is necessary to select an appropriate transducer to convert a mechanical quantity into an electric signal. The ratio between frequencies of natural vibrations of the transducer and the frequencies of vibrations

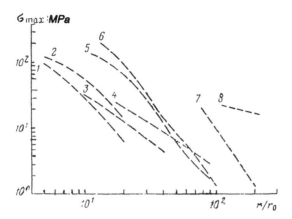

Fig. 3. Relationship between maximum stress σ_{max} and relative distance.

induced by the force acting is the decisive parameter in selection of a particular transducer under dynamic loads. The duration of a pulse of a compressional wave induced by blasting a group of microcharges in a model has been established at 180–360 μs and for a single charge 70–180 μs. It is obvious that recording of a given pulse is not possible by transducers having a frequency of natural vibrations less than 1000 Hz (contact type, inductive type and capacitor type). In such conditions, magnetoelastic and piezoelectric transducers with a frequency of natural vibrations of the order of several tens of kHz are more suitable. With such transducers, the recorded mechanical quantity should act directly on either the core of the magnetoelastic element or the piezoelectric element. Hysteresis of the magnetoelastic element is about 5–7% and the two elements are not interchangeable. The piezoelectric quartz transducer with longitudinal or transverse effect can be made to have frequencies of natural vibrations of up to 50 kHz. It is possible therefore to use it for recording high-frequency mechanical quantities.

Stresses induced by a blast in water are recorded by piezoelectric transducers in accordance with the method applicable to the medium (water). This method enables one to record only the compressive stresses since due to the cavitation of water at the rock-water interface, tensile stresses are not registered by the transducers. But in underground mining practice, data on tensile stresses originating at the interface of the ore fill material are very essential for controlling the blast effect.

In such conditions, the *method of measuring deformations by wire strain gauge* is of particular interest. However, this method cannot be used for measuring deformations in a rock massif subjected to blasting. Moreover, the resolution of such instruments is inadequate for recording the pulse of a blast wave. In this context the direct method of recording stresses induced by a blast in a hard medium is of significant interest. The merit of this method was also noted by R.M. Davis.

It is well known that the effect of blasting in ores is accompanied by displacement (deformation) of host rocks governed by the laws of propagation of blast waves and the nature of attenuation of waves with an increase in distance from the blast site. The effect of a blast pulse on the stability of the massif beyond the face is denoted by

$$I = \int_{t_0}^{t_1} \sigma \, dt,$$

where σ—blast-induced stresses. The construction of any transducer depends on the specific task of any investigation. In modelling of rock breakage by blasting, the transducer senses loads when the waves of compression as well as the tensile waves reflected from the boundary pass through. It is necessary to protect the strain gauge elements from mechanical effects and humidity to ensure their stable position in the model block and to make their retrieval possible for subsequent experiments. The transducers should be mutually interchangeable.

The gauge recommended by the author is suitable for recording stresses induced by a blast in a hard medium. The concept of possible usage of an elastic rod with uniform resistance for measuring strains under longitudinal compression or tension forms the basis for constructing such a transducer. If a steel rod is installed in a model block made of equivalent material at a distance b from charge 1, then the blast-induced force P acts on strain gauge 2, as a result of which the rod deforms longitudinally. The extent of deformation can be established with the help of wire strain elements glued to the rod. When in any section of the rod the stress is acting in a particular direction and strain is measured in the same direction, then Hooke's law, which establishes a constant ratio between stress σ and strain ε, is applicable:

$$E = \sigma/\varepsilon. \tag{1.25}$$

where E—elasticity modulus.

The length I and cross-section S of a wire as well as its specific resistance ρ vary during deformation of the rod and strain element:

$$R = \rho I/S.$$

Deformation of the strain element is accompanied by a proportional change in electrical resistance in the bridge circuit. The coefficient of sensitivity of wire to elongation (shortening) is the ratio of relative change of resistance $\Delta R/R$ to relative elongation ε. This is expressed in a differen-

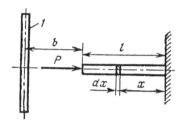

Fig. 4. Installation layout of charge and transducer

tial form by A.M. Turichin [33]

$$k = \frac{1}{R}\frac{\Delta R}{d\varepsilon} = (1 + 2\mu) + \frac{1}{\rho}\frac{d\rho}{d\varepsilon}. \qquad (1.26)$$

The sensitivity coefficient consists of a geometrical part based on Poisson's ratio μ, and a physical part—the coefficient of variation in specific resistance ρ during elongation. Values of k for wires used in manufacturing strain elements are as follows:

Constantan [Cu-Ni alloy]	1.8
Ni-Cr alloy	2.63
Platinum	4.12
Gold	3.22
Nickel	2–20

It is important to know the temperature-dependent variation in the resistance of wire. The temperature coefficients of resistance per 1°C for different wires are:

Constantan	0.00001
Ni-Cr alloy	0.0004
Nickel	0.006
Platinum	0.003

Taking into account the sensitivity coefficient and temperature coefficient of resistance, the authors selected strain elements made from constantan of the type 2PKB-20-200KhB. These elements possess a resistance of 205, 50–205, 99 ohm; the sensitivity coefficient at $+20 \pm 5°C$ is 2.26 (error 0.2%); relative elongation ± 3000 μm/m; and operating current not exceeding 14 mA. After glueing the strain element to the rod, its resistance changes by 5–9%. Readings of selected strain elements do not vary at temperatures from $+70$ to $-40°C$, which ensures recording of stresses with adequate accuracy.

The blast wave pulse is recorded on an apparatus containing a gauge, bridge, blasting device, oscillograph, and d.c. supply.

Working gauge: The main condition that prompted the use of wire strain elements in the modelling of breakage by blasting is their stable positioning and specific orientation in the model block. With this aim in view, a steel rod of uniform resistance was chosen and installed in the model block perpendicular to the axis of the drillhole charge. Four strain elements linked in series (pairwise from each side) were attached with tsiacrin 'EO' ($C_6H_7NO_2$) glue to the surface of the rod. The working gauge (Fig. 5) is a rod of rectangular cross-section with a die of 20 mm diameter. The die (surface adjoining the surrounding medium) takes the load from the blast.

Length of gauge, mm ...	108	138	168	198	228
Weight, kg ...	0.72	0.91	1.1	1.27	1.46

The rod with the lower die serves as the casing bottom, in which an aperture is left for inserting a cable (RK-75). One of the output terminals of converters is connected to the central wire of this cable and the other with its screen. Strain elements are affixed to the rod according to the guidelines laid down in TU PVYa 453-66 on attaching strain elements with tsiacrin 'EO' glue. Affixing strain elements to both surfaces of the rod ensures registration of deformations only due to compression or tension. When the rod is subjected to bending, two strain elements work on tension while the remaining two on compression. Thus the occurrence of bending phenomena is ruled out and the signal given is due only to longitudinal deformation because the strain elements are connected in series.

The main advantage of such transducers is that the sensitivity of the transducer itself increases when the resistance of the strain elements affixed to the rod increases. The increase in sensitivity of transducer through

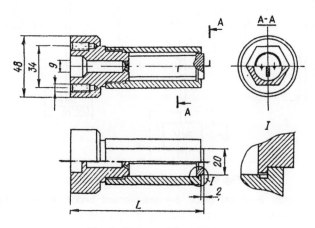

Fig. 5. Design of the gauge.

electrical means is more effective as distinct from the effect achieved by varying its geometrical parameters. During investigations, the base of measurements was constant (20 mm) and inertia did not exceed several microseconds. The casing of the transducer is hollow and is capped for protecting strain elements from mechanical damage and humidity. The working part of the upper die of the rod protrudes out of the cap so that during loading adequate deformation is ensured for recording purposes. The base of the transducer is a circular flange with two holes for affixing it to the metallic stand of the model. The distance from the charge axis to the surface adjoining the upper die of the rod and surrounding hard medium is adjusted according to the scale of modelling and the change in length of the casing flange. Transducers of such a construction can be retrieved for repeated usage in subsequent blasts.

The electrical measurement circuit (Fig. 6) consists of the transducer inserted into an arm of a symmetrical bridge.

The oscillograph is connected to the diagonal arm in the bridge. The bridge is powered by d.c. at 24 V by means of alkaline batteries. The bridge circuit is used as it provides higher sensitivity and adequate accuracy in measurements. Each arm of the bridge consists of four wire strain converters connected in series, with a total resistance of 822.0–823.8 ohm. The working gauge is connected to one arm of the bridge which is installed in the model block, while a similar rod of

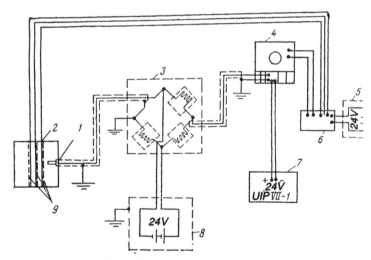

Fig. 6. Electrical measurement circuit

1—working gauge; 2—model block, 3—measurement bridge; 4—oscillograph, 5 and 8—power supply blocks; 6—block for blasting charges and for activating the oscillograph; 7—amplifier; 9—charges.

transducer with strain elements affixed to it is attached to the other arm. Two other contiguous arms of the bridge are also assembled from the same strain elements, which are glued from both sides to an elastic metallic diaphragm. With the help of a regulating screw, the plate is stretched at one arm and contracted at the other. In this manner the parameters of the bridge and output voltage in the diagonal arm are controlled. Using the regulating device, the bridge can be brought back to null position.

Under a blast the rod in the working transducer becomes deformed, causing a disturbance in the equilibrium of the bridge and consequently giving an output voltage in the diagonal. This output voltage is recorded by the oscillograph.

Blasting device (Fig. 7): Made on the basis of a selector switch (SS) of a telephone commutator with certain modifications, it ensures high accuracy of delay intervals in short-delay blasting. The sweeping cursor (SC) of the SS is installed on the axis of a synchronous electrical motor with a frequency of rotation 50 s^{-1}. The period of one turn of the cursor is equal to 20 ms. All the 50 contacts, which are radially situated, are switched on at one turn of the sweeping cursor, and switching time of each contact amounts to 0.4 ms.

The blasting device consists of a power supply block, SS with synchronous electric motor, combined switchboard panel (CSP), five commutator pins (CP), five output pins and one thyristor control.

The power supply block comprises an input transformer (IT) of 20–30 W and two rectifier bridges M1 and M2 with smoothening capacitance filters C1 and C2.

The output voltage of the device is realised through five thyristors (KY201) and is controlled by lamps LI-LV.

The electric motor is started by pressing the switch K1. By pressing the blocking switch BS, the thyristor KY-H is activated at the moment of closure of contacts 49 and 50 by the cursor. The output voltage KY-H is fed to SC. This initiates the sequential switching of the remaining 48 contacts of the SS, which are connected to the CSP. To select a particular delay interval, the controlling electrodes of thyristors KYI-KYV are connected to the CSP. A constant voltage is supplied at BI-BV by the opening of KYI-KYV. The required delay interval is prefixed by connecting the CP pins to the corresponding sockets of the CSP. The blasting circuit is fed by 24 V d.c.

The oscillograph should be such as to take into account the parameters of the effect of the blast wave on the gauge. The time parameter for model experiments is determined by the time of detonation of a 40-cm long charge (114×10^{-6} s) and time of expansion of detonation products ($100 \times 10^{-6} - 300 \times 10^{-6}$ s). Obviously, electromechanical oscillographs

20

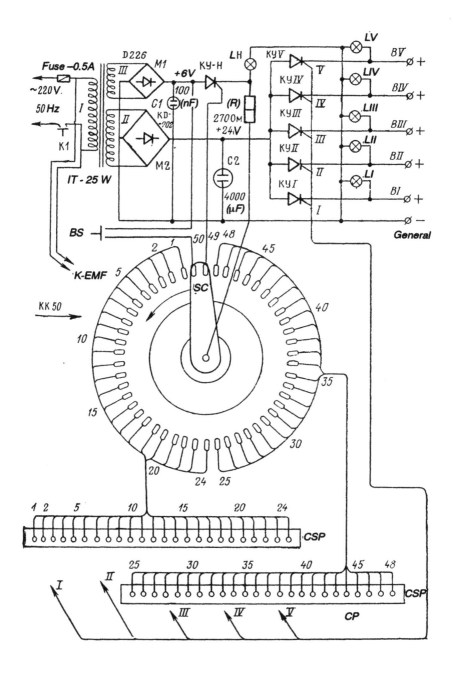

Fig. 7. Circuit of blasting device.

cannot be used for recording such a short-period electric signal. Moreover, they are inert and hence unsuitable for studying high-speed processes.

A balanced bridge of strain gauges fed by direct current is included in the measuring circuit. During the blast, the equilibrium of the bridge is disturbed and the output voltage on the oscillograph is seen as a beam. An oscillograph model C1-19B, equipped with d.c. amplifiers, was used in the investigations.

In the experiments, the blast wave pulse was recorded at a scanning rate of 50 μs/cm with calibrated pulses of 20 and 15 mm amplitude, as a result of which the entire process could be recorded. When it was required to record voltages using a five-beam oscillograph C1-33, the amplifier blocks of a.c. were replaced by blocks of symmetrical amplification of d.c. The oscillograph was fixed at a scanning velocity of 100 and 50 μs/cm with an amplitude of calibrated pulses 20 mm. Such oscillographs are non-inertial instruments, having a frequency characteristic in the range of 0–350 kHz. Taking into account the fact that maximum frequency of a working blast pulse in experiments was 5 kHz, it may be concluded that measurements were taken without significant errors. The power supply block of bridges and blasting circuit consisted of 24-V alkaline-type 2 batteries.

The first and second cascades of preamplifiers of the oscillograph C1-33 were powered by two double-channel universal feeders UIP-7-1 with a stabilised voltage of 48 V. Each power supply block had separate channels of stabilised voltage. This helped in eliminating interference.

The above method was developed for directly recording blast-induced stresses in a hard medium and the instruments used in experimentation were characterised by adequate resolution capability. This methodology is unique among known ones in that it is possible to record not only the compressive stresses, but also the tensile stresses. This capability enlarges the field of application of the method of equivalent materials.

The relationship between stress and relative distance in a blast in equivalent material is shown in Fig. 8. This functional relationship is of the type

$$\sigma_{max_m} = 800/(r/r_0)^{1.1}.$$

Furthermore, the interrelation between σ_{max} and relative distance in a blast in rocks as established by Borovikov and Vanyagin [7] is well known:

$$\sigma_{max_a} = 7 \times 10^4/(r/r_0)^{1.1}.$$

The comparative evaluation, using the above formulae, showed that at a relative distance of 40, stress in the model was 14 MPa, while under actual conditions it was found to be 133 MPa; when resistance of material against failure in real life was 150 MPa, in the model it was 1.8 MPa. The

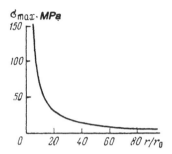

Fig. 8. Relationship between the maximum stress induced by a blast of TEN in equivalent material and the relative distance

interrelation between stresses in the model and the actual corresponded to the conditions of similarity (1.15).

1.5 Characteristics of Strain Gauges Having a Rod of Equal Resistance

In a model block in which a working gauge is installed, stresses are induced by a blast. In order to quantify these stresses, data on calibration of gauges are required. To use the data on static calibration of strain gauges in investigations of dynamic processes, it is essential to develop conditions of their application. As early as 1950, A.N. Krilov, while studying a mining shaft, observed: 'In order that a time-dependent force acting on any elastic system be considered as a static effect, from the point of view of deformations and stresses, it is essential that the period of natural vibrations of the system, and if such periods are many, then the largest among them, be less than the duration of force build-up from its zero value to the largest'. Ya.G. Panovko established that in order to eliminate the resonance effects of a strain gauge under dynamic loads, it is essential that natural frequencies of longitudinal vibrations be larger than the frequencies of vibrations of the acting force.

The law of conservation of mechanical energy is used to calculate natural frequencies of longitudinal vibrations of the gauge rod and the strain element simultaneously. In this case, the sum of kinetic and potential energy in the system does not change with time:

$$T_k + U_{pe} = \text{const,} \tag{1.27}$$

where T_k—kinetic energy of rod vibrations;

U_{pe}—potential energy of elastic deformation in rod.

Let us isolate at a certain distance x, a segment dx of rod (see Fig. 4). It is known that the longitudinal displacement of a rectangular rod, within the limits of Hooke's law, is determined by the ratio

$$\varepsilon = Pl/(EF).\qquad(1.28)$$

where P—force acting on the rod by blasting a charge;
l—length of rod;
E—elasticity modulus;
F—cross-sectional area of the rod.
    The relative rigidity (coefficient of rigidity) is expressed by the ratio of rigidity EF to rod length:

$$P/\varepsilon = EF/l = k.\qquad(1.29)$$

    As is known, vibrations at small deformations follow a harmonic law. Displacement of the rod at a distance x from the point of fixation would be

$$\varepsilon = \frac{Px}{EF}\sin\omega t.\qquad(1.30)$$

It is also known that the kinetic energy of the rod is

$$T_k = mv_k^2/2.\qquad(1.31)$$

Velocity $\qquad v_k = d\varepsilon/dt = Px/(EF)\,\omega\cos\omega t.\qquad(1.32)$

The kinetic energy of an elementary mass is calculated by

$$dm = \gamma_0 Fdx/g.\qquad(1.33)$$

where γ_0—density; g—acceleration due to gravity.
    Taking into account (1.32) and (1.33), we obtain

$$dT_k = \frac{\gamma_0 Fdx}{2g}\left(\frac{Px}{EF}\right)^2\omega^2\cos^2\omega t.\qquad(1.34)$$

The kinetic energy of rod vibrations is determined by integrating (1.34)

$$T_k = \int_0^l \frac{\gamma_0 F}{2g}\frac{P^2 x^2}{E^2 F^2}\omega^2\cos^2\omega t = \frac{\gamma_0 l^3}{6g}\frac{P^2}{E^2 F}\omega^2\cos^2\omega t.\qquad(1.35)$$

The variation in potential energy of elastic deformation in the rod is expressed by the relationship

$$U_{pe} = \frac{k\varepsilon^2}{2}\sin^2\omega t.\qquad(1.36)$$

    Substituting the values of ε and k from (1.28) and (1.29) in (1.36), we obtain

$$U_{pe} = \frac{P^2 l}{2EF}\sin^2\omega t.\qquad(1.37)$$

Replacing T_k and U_{pe} by their values from (1.35) and (1.37), the expression for law of conservation of energy (1.27) can be written as

$$\frac{\gamma_0 l}{6g} \frac{P^2}{E^2 F} \omega^2 \cos^2 \omega t + \frac{P^2 l}{2EF} \sin^2 \omega t = \text{const.} \qquad (1.38)$$

The process of rod (gauge) vibrations is accompanied by transition of potential energy into kinetic and vice versa. Let us consider the boundary conditions for these vibrations

I. $\omega t = 0$:

$$\sin \omega t = 0, \cos \omega t = 1,$$

therefore

$$\frac{\gamma_0 l^3}{6g} \frac{P^2}{E^2 F} \omega^2 = \text{const,} \qquad (1.39)$$

i.e., potential energy is converted totally into kinetic.

II. $\omega t = \dfrac{\pi}{2}$:

$$\sin \omega t = 1, \cos \omega t = 0,$$

we obtain

$$\frac{\gamma_0 l^3}{6g} \frac{P^2}{E^2 F} \omega^2 = \frac{P^2 l}{2EF} = \text{const,} \qquad (1.40)$$

i.e., kinetic energy is converted totally into potential.

Taking into account (1.39) and (1.40), we obtain

$$\frac{\gamma_0 l^3}{6g} \frac{P^2}{E^2 F} \omega^2 = \frac{P^2 l}{2EF} = \text{const.} \qquad (1.41)$$

From (1.41), the angular frequency of vibrations

$$\omega = \frac{\sqrt{3Eg/\gamma_0}}{l}. \qquad (1.42)$$

The frequency (Hz) of longitudinal vibrations of rod can be determined by

$$\nu = \frac{\omega}{2\pi} \frac{\sqrt{3Eg/\gamma_0}}{2\pi l}. \qquad (1.43)$$

The experimentally established frequencies of natural vibrations of a 2-mm thick rod are about 17.5×10^3 Hz. Compared to the calculated data, as per formula (1.43), the deviation is about 1.1%. In this case, the elasticity modulus for steel rods was determined using the ultrasonic device IPA-4 ($E = 0.98 \times 10^6$). If the tabulated data for steels ($E = 2.1 \times 10^6$) are taken, the calculated value of frequency of natural vibrations would be more.

The pulse of a blast wave is determined by the variable force applied to the gauge and the duration of its effect. If the least duration t of the effect of a variable force on the gauge in blasting a group of TEN microcharges in equivalent material is equal to 200×10^{-6} s, the maximum frequency of vibrations of the acting pulse is given by

$$\nu_p = 1/t = 5 \times 10^3.$$

In this case, the ratio of frequency of natural vibrations of gauge to the frequencies of vibrations of acting pulse $\nu/\nu_p = 3.5$. This justifies the use of strain gauges in recording blast-wave parameters.

The necessity for obtaining initial data quantitatively and enhancing the reliability of measurements is linked to the calibration of working gauges. Quantitative data on calibration of individual gauges help in explaining the effect of various factors on the results of measurements and in minimising errors. Thus if the gauge is subjected to compression, the relative change in resistance of strain element initially occurs according to the broken line 2 in Fig. 9 with angular coefficient $\tan \alpha$. After attaining a certain value of relative shortening, the gradient gradually reduces. In the subsequent reduction of stress on the gauge rod, relative variation in resistance also decreases according to the broken line 2. In repeated loading of the gauge, the resistance varies as per line 3 (Fig. 9). In subsequent cycles of loading and unloading of the gauge, the curves showing change in resistance of strain element approach each other and finally merge into line 1, i.e., mechanical hysteresis of the strain element disappears.

Calibrated gauges are used in experiments conducted for recording stresses induced by a blast. The layout for calibrating a gauge on the universal mechanical machine UMM-5 is shown in Fig. 10 (indicated by 2 in the Figure). When the gauge (1) is subjected to loading, the rod and strain elements attached to it also become deformed. The extent of deformation of the rod is converted, with the help of the measuring bridge (3), into an electrical signal and is recorded by the oscillograph (5). Calibrated gauges having rods of equal resistance have a constant characteristic: A linear relationship exists between the applied load and deformation of the rod (Table 4) or, in other words, between the load on the gauge and the amplitude of the beam on the screen of the oscillograph. The oscillograph and measurement bridge are powered by blocks 4 and 6.

The maximum load on the gauge is determined from the ratio

$$P_{max} = l_{st} P_k / l_k, \tag{1.44}$$

where l_{st}—amplitude of signal under study on the oscillogram, mm;

P_k—calibrated load, N;

l_k—amplitude of calibrated signal on the oscillogram, mm.

The maximum stress developed in a hard medium (at the interface of gauge/hard medium) is determined by

$$\sigma_{max} = P_{max}/S, \tag{1.45}$$

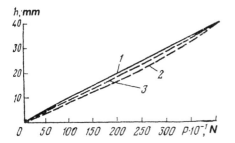

Fig. 9. Approximate relationship between amplitude of the beam on an oscillograph screen and load during calibration of the gauge (total resistance of gauge, 800 ohm).

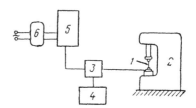

Fig. 10. Layout for calibrating a gauge.

Table 4

Load on gauge, N	Amplitude of calibrated signal, mm	Load on gauge, N	Amplitude of calibrated signal, mm
Rod of 2-mm thickness		Rod of 4-mm thickness	
5	3.5	10	3.5
10	7 0	20	7 0
15	10 5	30	10 5
20	14.5	40	14.5
25	17.5	50	17 5
30	21 0	60	21.0
35	24 5	70	24 5
40	28.0	80	28 0

where S—area of gauge surface in contact with hard medium, mm².

Data from experimental investigations show that wire strain gauges possess the following major advantages: capability for responding under static and dynamic deformation and a high degree of sensitivity to deformation. Significantly, strain gauges happen to be almost non-inert. Yu.I. Iorish observed that the sensitivity of strain gauges does not vary even at deformations occurring under frequencies of 20,000 Hz if they are properly glued.

In modelling the breakage process due to blasting, when stresses were recorded at $r/r_0 = 30-40$, in almost all cases the strain element failed. This was due to the wire breaking at the point of its attachment to the outgoing conductor. The average number of loading cycles until failure of the strain elements is about 20–30.

Under dynamic loads, in particular during a blast, to ensure reliable working of the gauges the compressed rod is permitted to deform up to a limit such that its working is not affected. Therefore, it is essential to test the gauge rod for strength and stability. Experience has shown that at $r/r_0 = 30$ in a blast using TEN, the maximum stress is 22 MPa. As the relative distance is reduced, the intensity of stresses increases and at $r/r_0 = 10$ reaches 48–50 MPa. This condition presupposes that the rod is working on compression. Belyaev [6] observed that it is not always possible to damage the rod by bringing the compressive stresses to the yield stress level or the ultimate strength of material. The rod fails because it cannot maintain its original form of rectangular compressed unit and becomes distorted; this causes bending moments under compressive forces. As a result, additional stresses due to bending develop and the rod loses its stability. This has to be borne in mind while testing the compressed rod for compression and stability.

The measurable relative deformation of strain gauge 2 PKB-20-200-KhB serves as the base for calculations and does not exceed 3000 μm/m (0.003 cm/cm). The relative deformation of a rod with uniform resistance can be obtained from (1.25). Experiments for determining the elasticity modulus through observations of elastic vibrations of rods [6] have shown that even under dynamic loading, Hooke's law is valid and the elasticity modulus retains its value. As for the nature of the development of stresses and deformations, deformation occurs even though a blast may be very fast, but is not instantaneous. The known quantities in (1.25) are σ and E. Stress in the rod can be expressed as

$$\sigma = P_{max}/F. \qquad (1.46)$$

where P_{max}—maximum load on the gauge during the blast;
F—cross-sectional area of rod.

$$P_{max} = \sigma_{max}\pi d^2/4, \qquad (1.47)$$

where σ_{max}—maximum stress recorded at a known relative distance;
$\pi d^2/4$—gauge surface in contact with the hard medium (in experiments $d = 20$ mm).

At $r/r_0 = 30$, $P_{max} = 69$ N. Consequently, the stress in the rod with a cross-sectional area of 48 mm^2 is equal to 144 MPa.

Using (1.25), relative deformation of the rod is determined (0.00147 cm/cm). The absolute shortening of the rod is found to be (0.0079 cm):

$$\Delta l = \varepsilon l. \qquad (1.48)$$

where l—length of rod, equal to 5.4 cm.

The selected cross-section should satisfy the strength condition

$$\sigma_{per} \leqslant |\sigma_{per}|. \qquad (1.49)$$

where $|\sigma_{per}|$— permissible value of normal stresses during blast, determined according to

$$|\sigma_{per}| = \sigma_*/k_{fos}, \qquad (1.50)$$

where σ_*—yield limit of rod, equal to 400 MPa;
k_{fos}—factor of safety.

In (1.50), the factor of safety under dynamic loading is not known. Its value is determined according to the following formula [6]:

$$k_{fos} = 1 + \sqrt{1 + \frac{2T_k l}{\Delta l^2 EF}}, \qquad (1.51)$$

where T_k—kinetic energy of the blast. Assuming that the kinetic energy converts entirely into potential energy of elastic deformation in the rod, we can write,

$$T_k = U_{pe}. \qquad (1.52)$$

The potential energy of elastic deformation in the rod is given by

$$U_{pe} = P_{max}\Delta l/2. \qquad (1.53)$$

By substituting the results of computations from (1.52) and (1.53) in formula (1.51), we get $k_{fos} = 2.42$. The permissible normal stress in the rod, as obtained from (1.50), is equal to 165 MPa. It is to be noted that the actual stress (144 MPa) in a rod of cross-section 0.3×1.6 cm, is less than permissible.

In the case of inadequate thickness of the rod, its stability is disturbed. Stability of a compressed rod is determined by the condition

$$\sigma_{cal} \leqslant |\sigma_y|, \qquad (1.54)$$

where σ_{cal}—calculated stress

$$|\sigma_y| = \varphi|\sigma_{per}|, \qquad (1.55)$$

where φ—coefficient of reduction of principal permissible stress for compressed rods. Under condition (1.54) and relationship (1.55), the value of $|\sigma_{per}|$ is known; unknown quantities are σ_{cal} and φ.

Let us determine the moment of inertia of a compressed rectangular rod, relative to the neutral axis oz (Fig. 11)

$$I_{min} = \int_{-b/2}^{+b/2} hy^2 dy = h\left[\frac{y^3}{3}\right]_{-b/2}^{+b/2} = \frac{hb^3}{12}, \qquad (1.56)$$

where h—length of rod, equal to 1.6 cm;
b—thickness of rod, equal to 0.3 cm.

From (1.56), the moment of inertia is found to be 0.0036.

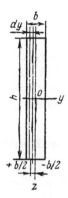

Fig. 11. Diagram for computing moment of inertia of a compressed rod.

It is evident from the course of strength of materials that a major role is played by the slenderness of the rod in its resistance to buckling.

The minimum radius of inertia of section is given by

$$i_{min} = \sqrt{I_{min}/F}. \tag{1.57}$$

Substituting the values of I_{min} and F in formula (1.57), $i_{min} = 0.087$ cm. The rod slenderness is given by

$$\lambda = \mu_1 l/i, \tag{1.58}$$

where μ_1—length coefficient (0.725) is taken from the tables [6]. Slenderness of the rod, determined by formula (1.58), equals 45.

By interpolating the values of φ for values of λ between 40 and 50, we obtain $\varphi = 0.85$ [6].

The calculated stress at $(r/r_0) = 30$, would be

$$\sigma_{cal} = P_{max}/(\varphi F) = 169 \text{ MPa}.$$

The magnitude of overstress: $\dfrac{169 - 165}{165} \times 100 = 2.6\%$, i.e., $|\sigma_y| < \sigma$.

The blast wave pulse can be stably recorded while recording the stresses induced by blasting a charge of TEN in equivalent material at $r/r_0 = 30$ with a gauge bearing a 0.3-cm thick rod. Dimensions of cross-sections of other rods, which enhance the reliability of recording stresses at corresponding relative distances from the blast site (of a charge of TEN), are similarly determined. Results of these computations are given below.

r/r_0	10	20	30	40
σ_{max}, MPa	48–50	32–34	20–22	15–17

P_{max}, N	150–157	100–107	63–69	47–54
Cross-sectional area of				
gauge rod, mm	16 × 5	16 × 4	16 × 3	16 × 2

Stresses at $r/r_0 = 40$ can be recorded by gauges bearing a rod with a cross-section of 1.6 × 0.2 cm. A comparison of computed results with experimental data confirmed the validity of using the suggested method for recording stresses during modelling of rock breakage by blasting.

1.6 Methods for Recording the Growth of Cracks during Rock Breakage by Blasting

The processes of growth and branching of cracks have not been detailed in studies of the mechanics of brittle failure. This omission is ascribed to the difficulties in conducting experiments. Amongst the existing methods of experimental investigation of failure, use of the photorecorder SFR-1 is the most common. This method enables one to obtain photographs with a speed of 2.2 million snaps/s.

The method of recording cracks during breakage of acrylic plastic by blasting is well known. This involves illuminating the transparent material under study by means of a pulse of a luminous flux, blasting a charge and recording the process with synchronous high-speed photography. In this case, illumination and photography are done from directions opposite to the object under study. However, the application of this method in recording cracks during blasting in rocks is limited.

The surface of a prepolished (to a glassy condition) metallic specimen is illuminated by a pulse of luminous flux; the specimen is broken by means of blasting and the entire process is photographed simultaneously. The demerits in this method are: worsening of optical resolution caused by reflection of light flux from the glassy surface of the specimen; complexity of the process linked with polishing of the specimen surface until it attains a glassy state and synchronisation of pulse illumination with the blasting process. Furthermore, due to the glassy condition of the specimen, the SFR has to be installed at an angle relative to the surface of the specimen; therefore, it is not possible to establish the true extent of widening of the surfaces of the growing crack. These disadvantages notwithstanding, usage of this method has enabled establishment of the peculiarities involved in the initiation, growth and branching of cracks in metals under different loading conditions (Table 5). In the case of high-carbon steel ShKh 15 specimens, rupture was particularly intensive under heavy stresses. If the average rate of failure for a tempered specimen under a no-load condition was less than 1000 m/s, under a tensile stress of 300 MPa it was found to be 2300 m/s. The average rate of propagation of crack failure of photochromic glass under a stress of 30 MPa was equal to 1500 m/s.

At the impact testing stage by a pendulum type machine MK-30 with an energy content of 300 J, the rate of crack propagation was found to be 4–8 m/s. According to the investigations of V.M. Finkel, the largest known rates of failure are approximated to a velocity amounting to 0.38 of the velocity of longitudinal elastic waves in steel (see Table 5).

Another method of recording cracks is also known. On a 4-mm thick glass plate of 100×100 mm, rings of 0.8-mm wide silver paste are applied at a distance of 8–10 mm from each other. These rings are connected to the electrical circuit of a balanced bridge and also to the source of d.c. power. When a ring is ruptured, the bridge goes out of balance and voltage at the output mode is recorded by the oscillograph (cathode-type) OK-17.

The oscillograph is activated by the piezoelectric transducer, which is placed directly on the explosive charge. The time of rupture of the rings is established based on the oscillograms obtained. The velocity of propagation of cracks is determined according to the known distance between adjacent rings and the time of their rupture. In this case, the failure sequence of the rings may possibly be disturbed and the rate of movement of the failure front can be 1.5–2.5 times more than the rate of movement of the crack. This is related to the lead in failure, caused by the presence in materials of 'weak spot' cracks. Thus, for example, the rates of movement of cracks in rocks are given in Table 6. The rates of movement of the failure front are relatively more, as can be seen from Table 7.

An experimental investigation into the breakage of various sections of a bench 16–17 m in height by means of drillhole charges of 214 and

Table 5

Author	Material	Ultimate velocity of crack movement v_T, m/s	Sonic velocity C_p, m/s	v_T/C_p
V.M. Finkel	Steel	1500	5850	0 256
I A. Kutkin	Monocrystals of rock salt	2400	4500	0.536
	Acrylic plastic	400	2300	0.174
	Tempered steel	2300	5850	0.393
E. Field	Monocrystals of diamond	4400	12000	0.367
C Shardin and V. Street	Glass	1500	5180	0 289
V. Akita and K. Ikeda	Steel	2050	5580	0.367
E. de Noger and C Pollack	Glass	4000	5500	0.227
E. Gilman, J. Knudsen and W Walsh	Monocrystals	2000	6500	0.309

Table 6

Author	Material	Rate of crack movement v_T, m/s	Velocity of sonic wave C_p, m/s	v_T/C_p
M.F Drukovanyı	Granites	130–230	5500–6000	0.024–0.038
V M. Komir	Diabase	245	6260	0 039
	Hornfels	81–101	4350–4780	0.019–0.021
	Shales	580–670	5450–6310	0 106
N U. Turuta and A.T. Galimullin	Limestones	700–4000	5160–5875	0.136–0.681
G.N. Kuznetsov	Equivalent material based on sand	71	1630	0.043

230 mm is of particular interest. In these experiments wire strain gauges were used. The time intervals between initiation of charge detonation and the moment of failure of gauges were recorded on oscillograph H-700. In the course of this experiment, the variable rate of growth of crack at different points in the bench slope was found to be 405–630 m/s, at the top of the bench (between drill hole and bench edge) 204–240 m/s, while between charges in a row 58–170 m/s. During the breakage of boulders of dolomitic limestone by a charge of 28-mm diameter, the rate of growth of crack varied from 700 to 4000 m/s when the relative distance was varied between 16–36. This indicates the effect of breakage conditions in the dynamics of crack growth.

Summarising the results of experiments of breakage of materials, it can be seen that the total duration of exposure of frames using SFR-1 did not exceed 100 μs, which is far less than the time taken for failure of the medium itself. In recent years SFR-2 has been used, which is capable of exposing frames up to 500 μs. However, even this time period is insufficient for recording the process of rock breakage by blasting. Moreover,

Table 7

Author	Rock	Rate of failure of massif, m/s
M.F Drukovanyı	Limestones	1200–1310
V.M. Komir	Granites	700–800
	Diabase	1000–1200
	Hornfels	331–392
	Shales	985–1050
	Dolomites	1040–1380
N.U. Turuta and	Limestones	—
A.T. Galimullin	Dolomites	405–630

because of their special properties, the usage of optically active materials (epoxy, polymers etc.) does not ensure attaining mechanical similarity between materials used in the model and the actual conditions.

Establishing the rate of movement of a single crack is inadequate for assessing the condition of the massif being broken. During breakage by blasting a high rate of movement of several cracks might develop; yet even then the ore (rock) may not get separated from the massif and fragmentation may not take place at all. This is due to inadequate movement and absence of branching of cracks. Thus there is a need to widen the existing field of investigation of rock-breakage parameters.

In metal mines, rocks or ores are blasted by group charges. Due to their interaction, an experimental study of the features of crack formation, is difficult. Usually the kinematics of cracks varies with change in the relative distance r/r_0. Taking this into consideration, the scheme of breakage and recording of cracks can be simplified by considering the working of a unit charge in the zone of active breakage. In such a case, the process of entire breakage of the model was found to be prolonged for 200–1000 μs, while the wave pulse existed for only 25–300 μs.

Under such circumstances, the need arose for developing a new method for recording the growth of cracks during a blast using the time loop LV-04 [3]. LV-04 is based on electronic-optical converters (EOP). Investigations of high-speed processes in plasma physics have been conducted using EOP. Specific examples of the use of EOP in investigations into the failure of optically inactive materials—rocks, concrete etc.—are not cited in the published literature.

LV-04 can work in two modes—frame type and linear. Separate frames are taken after 10, 100, 300 and 500 μs. Each frame is exposed for 10, 20 and 100 μs. It is possible to record clearly and completely the formation, growth and branching of cracks during a blast conducted in a block of materials, by establishing the corresponding exposure time and time interval between snaps. The time resolution attained in this mode is 10^{-7} s. The resolution ability of the experimental set-up is not less than 18 lines/mm for EOP of the UMI-92 type and not less than 15 pairs of strokes/mm for UMI-93. The time between switching on the instrument and putting it off does not exceed 10^{-6} s. A Zenith-5 camera equipped with a lens R-Helios 100 is used to photograph the image from the screen of the EOP. The image is focussed on the EOP by an MIR-1 lens. The instrument is supplied with 220 V and 50 Hz electric current. However, the resolution ability of the LV-04 does not permit simultaneous recording of the growth of blast-induced cracks in rocks. The well-known non-destructive methods for observing cracks in hard bodies based on fluorescence are not suitable for recording movement of blast-induced cracks in rocks.

This objective can be achieved in the following manner: in the method of recording failure of hard bodies, involving blasting with synchronised high-speed photography, a layer of luminescent material is laid on the top of the object under study prior to breakage and is activated before the blast by a pulse of light flux.

The block diagram of LV-04 is shown in Fig. 12. The principal scheme of the apparatus for recording the growth of cracks during breakage of materials by blasting (Fig. 13) includes: Object under study (1), source of light pulses (2), time loop (3), blasting device (4), and control block (5). In a rock block (1), a hole (6) of 3–5-mm diameter is drilled and on one of its sides luminescent material (solution of white phosphor with transformer oil in 1 : 2 ratio) is applied. Two charges of TEN (8) fusehead (9) with a drop of lead azide and stemming material (10) are placed in the drill hole.

The sample or block to be broken is placed in a portable blasting chamber and its surface (7) is activated by the pulse of light flux; blasting is then carried out synchronously with high-speed photography of cracks in darkness. The use of the synchronising device in this apparatus enables one to fix the moment of charge detonation and initiation of crack formation on the surface of the rock specimen.

The experimental installation includes: radio racks Nos. 1 and 2, container for electronic-optical convertor with an adjusting device and a trolley, control panel, trolley with voltage stabilisers SN 0.75, Zenith-5 camera with an attachment, lenses TAIR-3, RO2-2M, MIR-1, R-Helios 100, telescopic microscope, devices for synchronisation and blasting, oscillograph C1-65, universal power source UIP-7 and a portable blast chamber with a pulse lamp.

The patterns of crack formation in blocks of materials were studied using an LV-04. Blocks were placed on a clamping device in the blast chamber. The block massif being broken by blasting can only move in the direction of compensating cavity.

In rock breakage by blasting, it is important to establish the direction of movement of the failure front. This aspect is not dealt with adequately in the published literature. Hence to study the direction of movement of failure surfaces, it became necessary to conduct additional experiments as per the scheme illustrated in Fig. 14. Graphite rods of 2.3-mm diameter and 60-mm long were insulated with nitrocellulose lacquer and then installed in blocks of test materials and connected to the electrical circuit. During the blast macroscopic cracks formed and the rods broke simultaneously, as a result of which beams appeared on the oscillograph screen. These beams were photographed by the camera. The sequence of appearance of the beams indicated the direction of crack movement. According to the established duration of failure of the rods and also the

35

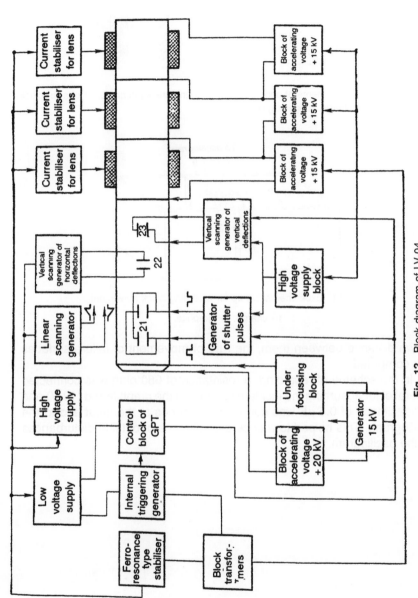

Fig. 12. Block diagram of LV-04.

21 — Shutter gate   22 — Filter of horizontal deflections   23 — Filter of vertical deflections

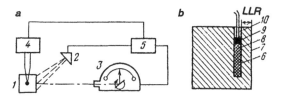

Fig. 13. (a) Principal scheme of apparatus for recording cracks during blast of a (b) block with an explosive charge. (Legend given in text.)

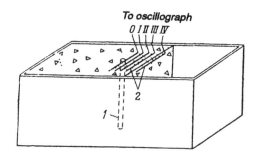

Fig. 14. Experimental set-up for recording crack movement.

1—charge; 2—graphite.

known distance between them, the rate of movement of failure surfaces was computed.

In the scheme of recording, a resistance of 680 ohm was connected in parallel for absorbing the incident voltage on the graphite rod (12 V) and a capacitance of 0.5 µF was connected in series at the input end to the oscillograph. This helped to reduce the influence of outside disturbances on the oscillogram.

2

Kinetics of Rock Breakage by Blasting

2.1 Rate of Crack Movement and Widening of their Surfaces

The results of investigations into the dynamics of growth of brittle cracks during a blast in rocks, concrete and equivalent materials are discussed below. Experiments were conducted according to two schemes. In one scheme, the origin, growth and branching of cracks on the surface of the object under study were recorded using time loop LV-04. In the second scheme, the moment of failure of the massif or specimen was recorded by means of 60-mm long graphite rods of 2.3-mm diameter. These rods were priorly insulated with nitrocellulose lacquer and were placed in the model block in groups of 4 to 5 at different distances from the charge. The rate and direction of movement of the crack were determined based on the frames obtained from high-speed photography of crack formation. The process was controlled through oscillograms that showed the moments of breakage of the graphite rods. If the value of lag is taken as 1 μs for EOP UMI-92 or UMI-93 and also no lag for graphite rods, it follows that the combined investigations of rock breakage in the two experimental set-ups (see Section 1.6) improved the reliability of these experimental studies.

The physicomechanical properties of the tested materials are given in Table 8. In order to maintain the natural condition of the test blocks, they were sawn in the Armmramor (Armenian marble) plant itself while blocks of rock salt were cut in the 235-m level of Avanskii salt mine (Erevan city).

The specific procedure of exposing frames involved the following. The growth of cracks in a block of cemented fill $300 \times 200 \times 180$ mm with a 4-mm diameter and 70-mm deep drill hole was recorded. A charge of TEN weighing 50 mg and lead azide fusehead were inserted in the drill hole. The remaining part of the hole was filled with stemming material.

Prior to this, a coat of luminescent material was applied to the surface of the test object. After drying the luminescent material over a period of 4–5 hours, the test block was placed in the clamping device of the blast chamber. The luminescent layer was activated by a pulse of light flux and

Table 8

Material	Density, g/cm^3	Strength, MPa		Poisson's ratio	Elasticity modulus, 10^{-5} MPa	Velocity of longitudinal wave, m/s
		Under compression	Under tension			
Andesitic porphyry	2 7	135	—	0 22	0 44	4370
Quartzitic porphyry	2 7	152	—	—	0 51	—
Secondary quartzite	2.7	137	—	0.26	0.57	4600
Basalt	2 6	72	—	0 23	0 48	3500
Marble	2.7	87	—	0 24	0 72	4100
Rock salt	2.1	33	15	0.28	0 31	4100
Equivalent material	2.6	4	0.4	0 20	0.01	650
Tuff	2 1	9	0.7	0 20	0 05	2500
Cemented fill	1 7	6	—	0 20	0.07	2200

later the block was blasted while taking synchronous photographs of the test object (in darkness) with a high-speed camera.

A bright cloud of gaseous products of explosive detonation was noticed in the first frame though there were no cracks in it. A major crack appeared only in the second frame after 340 μs elapsed from the initiation of charge detonation. Later, branched cracks appeared—one in the third frame and two in the fourth frame.

According to G.I. Barenblatt, opposite surfaces of a crack merge smoothly at its contour. Under such conditions, the rate of widening of surfaces in a crack v_b is an important parameter in the dynamics of ore (rock) breakage. In reality, the formation of several cracks in the layer being broken cannot lead to the separation of ore from the massif and its fragmentation. To achieve this, sufficient movement of the ore along the line of least resistance (LLR) is required and is determined by the rate and time of widening of crack surfaces.

Within a period of 300 μs, the rate of widening of crack surfaces (v_b) increased to 2.3 m/s, i.e., the acceleration value was 7.6×10^3 m/s^2. In all the experiments, the acceleration of cracks was established, characterising the rate of change in their growth. The fragmentation was more intensive as cracks approached the blast site (with reduction in LLR). In this experiment, LLR was equal to 12 mm, charge weight 180 mg, diameter 2 mm and length 64 mm. The exposure time of a frame was 20 μs while the interval between two successive frames was 100 μs. Cracks were noticed in the third frame, after 260 μs from the moment of charge detonation. The rate of widening of crack surfaces was found to be 10–17 m/s.

In a relatively near-zone of the blast, for example at LLR 10 mm, v_b in basalt was 20.8 m/s. In this case, a charge weighing 450 mg, 2.6 mm in diameter and 70 mm long was used. Subsequently, v_b decreased from 20.8 to 14.8 m/s.

The failure intensity lessened with an increase in LLR. The marble block chipped at LLR of 28 mm when a charge of 2-mm diameter was blasted. In this case, the widening rate of the surface was lower and equal to 1.3 m/s.

In a block of red tuff, pulverisation took place at LLR of 29 mm. An 80-mm long charge of 1.8-mm diameter weighing 205 mg was used for the purpose. Development of cracks ahead was noticed in the central part of the exposed surface on the opposite side of the charge. The rate of widening of the major crack was 7.9 m/s.

The data from experimental investigations show that the rate of widening of crack surfaces (movement of massif) significantly influences the branching of cracks (crushing of massif). Actually, the dynamics of development of individual cracks does not entirely determine the dynamics of the crushing process of the massif. Cases are known in which cracks were accelerated with a rate equal to half the velocity of a Rayleigh wave and yet branching did not occur; in other words, the massif did not break.

Several blasts were conducted in blocks of rock salt, which initially were practically devoid of cracks. The breakage process of a rock salt block $310 \times 275 \times 265$ mm by blasting a charge of 71 mg (diameter 3 mm, length 15 mm, LLR 24 mm) was photographed. The process was recorded with exposure time of each frame of 20 μs and time interval between frames of 300 μs. Cracks were noticed in the second frame with rate of widening of major crack surface 7.8–9.4 m/s and of branched crack at 2.1 m/s.

The growth of cracks induced by a blast in blocks of tuff of the Erevandakertskiy deposit (Armenia) was photographically recorded. In this case, the blocks were broken under conditions of LLR 24 mm, charge weight 169 kg (diameter 1.8 mm, length 67 mm). The tuff pulverised more intensively than the rock salt. In rock salt, a blast-induced zone was formed, which was dark in colour and the branching of cracks was clearly seen in this zone. The major cracks were oriented mainly along the generating line (directrix) of the charge. The widening rate of the surfaces of the major cracks in red tuff was found to be 8.3–11.4 m/s, while that of the branched crack 4.2 m/s.

In a block of cemented fill, the initiation of cracks was observed in the fifth frame. As the process extended in time, the rate of widening of crack surfaces also increased. Thus, between 4–8 frames, it increased by 1.5 mm and between 4–14 by 2.1 mm. The distance between the surfaces of one of the branched cracks was 1.5 mm; however this did not conform

40

to the model and was established only in the photograph. The true value of widening of surfaces was established by dividing the amount of opening of a crack in the frame by the scale of the photograph (1 : 5). The widening rate of crack surfaces was determined by dividing the extent of crack opening by the time fixed between adjacent frames.

The computed values of v_b in the materials tested are presented in Table 9. From this table, the ratio between v_b and velocity of elastic wave (c_s) was computed and is shown in Table 10.

Experience has shown that the ratios between v_b and velocity of transverse wave tend to 0.004.

As an example, let us consider the pattern of crack development during breakage of basalt by blasting. As the relative distance is reduced from 7.7 to 20 sic, i.e., by 2.6 times, the widening rate of crack surfaces increased by approximately 4.5 times (Fig. 15).

The experimentally obtained relationship between v_b and r/r_0 enabled establishment of the ratio between these two quantities as

$$v_b = \frac{600}{(r/r_0)^{1\,6}}. \qquad (2.1)$$

As the LLR decreased, v_b attained a value of three orders and the rock was intensively pulverised.

During the experimental studies, the need arose for conducting special experiments according to the scheme shown in Fig. 14 (see Section 1.6). Results of these experiments were required for verifying the direction of crack movement and for establishing the rates of movement of failure surfaces.

In the experiments, the material was broken in the direction of a compensating cavity. Concrete blocks were cast in a metallic box 400 × 200 × 200 mm with 15-mm thick walls. In experiments using rock blocks, the gaps between box walls and rock were filled with concrete mix.

Rock was blasted by charges of TEN with a charge density of 1 g/cm^3. The charge was placed in the block in the following manner. Before filling

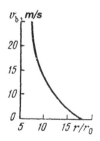

Fig. 15. Dependence of rate of widening of crack surfaces v_b on relative distance r/r_0 in fragmentation of blocks of basalt by blasting.

Table 9

Medium (1)	LLR, mm (2)	Charge weight (TEN), mg (3)	Exposure time of frame, μs (4)	Time interval between frames, μs (5)	No. of frames (in numerator) and distance between surfaces of cracks, mm (in denominator) (6)	Extent of widening of surfaces of cracks (mm); on the frame in the model (7)	Duration of widening, μs (8)	v_b, m/s (9)	Crack type (10)
Basalt	10	450	20	100	3/0.5; 6/2	1 5/7 5	360	20.8	M*
	10	450	20	100	3/0 5; 10/3	2.5/12 5	840	14 8	M
	16	180	20	100	4/0.8; 9/15	0 7/3.5	640	5.4	M
	10	180	20	100	3/0 5; 5/1.1	0.6/3	240	12.6	M
	20	190	20	300	3/1; .7/1 8	0.8/4	1280	3 1	B**
	20	190	20	300	4/1.1; 8/2.5	1.4/7	1280	5.4	M
	20	190	20	300	3/0.9; 13/4	3.1/15.5	3200	5.0	B
	12	180	20	100	6/0.8; 10/1.8	1/5	480	10 4	M
	12	180	20	100	10/1.8, 13/3	1 2/6	360	16 6	M
	13	176	20	100	7/1; 11/2	1/5	480	10.4	M
Marble	28	222	20	300	4/1 1; 16/2.1	1/5	3840	1.3	M
Tuff	25	124	20	100	2/0.4; 3/0.8	0 4/2	120	10 7	M
	25	124	20	100	5/1.0, 9/1.8	0 8/4	480	9.3	M
	25	124	20	100	10/2; 14/3 1	1.1/5.5	480	11.4	M
	25	124	20	100	7/0.7, 11/11	0.4/2	480	4 2	B
	24	169	20	100	5/0.7; 9/1.9	1.2/6	480	12.5	M
	24	169	20	100	6/0.5; 10/1.1	0 6/3	480	6.2	B
	29	205	20	300	6/2; 10/4	2/10	1280	7.9	M
Rock salt	24	71	20	300	4/0.6; 7/1	0.4/2	960	2.1	B
	24	71	20	300	4/0.8; 6/1.8	1/5	640	7.8	M
	24	71	20	300	4/1; 6/2.2	1.2/6	640	9 4	M

Contd

(1)	(2)	(3)	(4)	(5)	(6)	(7)	(8)	(9)	(10)
Cemented fill	30	80	10	300	4/0 5, 8/2	1 5/7 5	1240	6 0	M
	30	80	10	300	4/0 5, 14/2 6	2.1/10 5	3100	3 4	M
	30	80	10	300	7/0 6, 15/2 1	1 5/7 5	2470	3 0	B
Equivalent material	30	80	20	100	7/0.5, 15/1 2	0 4/2	960	2 1	B
	30	80	20	100	6/0 5, 16/1 4	0.9/4 5	1200	3 7	M
	30	80	20	100	5/2, 16/3 1	1 1/5 5	1320	4 2	M
	30	80	20	100	7/0 7, 9/1.3	0.6/3	240	12.5	M
	30	80	20	100	10/0 4, 16/1	0.6/3	720	4 1	B
	30	80	20	300	5/0 6, 6/0 9	0.3/1 5	320	4 7	B
	30	80	20	300	3/1.5, 7/5	3.5/17 5	1230	13 7	M
	30	80	20	300	3/0 2, 5/0 8	0.6/3	640	4 7	B
	30	80	20	100	5/0 5, 9/1 2	0.7/3 5	480	7 2	M
	30	80	20	100	5/0 5, 16/1 5	1/5	1320	3 8	B

*M—Major
**B—Branched

43

Table 10

Medium	c_p, m/s	c_s, m/s	v_b, m/s	v_b/c_s
Basalt	3500	2240	5 4	0.0024
	3500	2240	5	0 0022
	3500	2240	10 4	0 0046
	3500	2240	16.6	0 0074
	3500	2240	20 8	0 0092
	3500	2240	14 8	0 0066
Tuff	2500	1550	10 7	0.0069
	2500	1550	8 3	0 0053
	2500	1550	11 4	0 0073
	2500	1550	12 5	0 008
		1550	6.2	0 0039
	2500	1550	7 9	0 0051
Rock salt	4100	3000	7.8	0 0026
	4100	3000	9 4	0.0031
Cemented fill	2200	1320	6	0 0045
	2200	1320	3 4	0.0026
Equivalent material	650	390	3 7	0 0091
	650	390	4 2	0.0107
	650	390	12 5	0 032
	650	390	13.7	0 0351
	650	390	7.2	0 0185
	650	390	5.7	0 015
	650	390	4.7	0 012
	650	390	15	0 0405

the box with concrete, a metallic rod of 5-mm diameter was placed in the metallic frame of the model. After the model attained the required strength, the rod was withdrawn and a charge enclosed in a glass tube was placed in that cavity. The charge was initiated by blasting the lead azide fusehead. Adequate rigidity of the system chosen prevented any lateral movement of the massif under study. The picture of movement of beams, as observed in the oscillogram, was the principal schematic of massif breakage. When the rod was damaged, the beam dropped due to a decrease in potential (voltage in the circuit). The experimental results are given in Table 11.

According to the moments of breakage of rods, as established from the oscillograms, and the known distance between rods, the rate of crack movement was determined. As the crack approached the exposed surface, prefailure in the form of chipping was distinctly visible. Chipping occurred in 20 out of 41 experiments conducted.

The change in rate of crack movement depending on the relative distance during blast in blocks of secondary quartzite, tuff, basalt and concrete, is shown in Fig. 16. An exponential function of the following

Table 11

Serial number	Medium	Compressive strength of material, MPa	LLR, mm	Charge diameter, mm	Charge length, mm	r, mm	r/r₀	Duration of loading until rod breakage, μs	vᵣ, m/s	v_T/C_P	Remarks
(1)	(2)	(3)	(4)	(5)	(6)	(7)	(8)	(9)	(10)	(11)	(12)
30	Concrete	20	50	2	155	7	7	200	–	–	
		20	50	2	155	21	21	300	280	0.14	
		20	50	2	155	31	31	380	125	0.06	
31	Concrete	6	100	2	172	28	28	425	–	–	
		6	100	2	172	73	73	1500	42	0.03	
32	Basalt	72	200	2.6	75	40	30.8	220	–	–	
		72	200	2.6	75	110	84.6	1970	40	0.01	
		72	200	2.6	75	180	138.4	–	–	0.01	
33	Basalt	72	100	3	18	19	12.6	20	–	–	Chipping
		72	100	3	18	38	25.3	–	–	–	
		72	100	3	18	57	38	320	127	0.04	
		72	100	3	18	76	50.6	180	–	–	
34	Tuff	9	100	2.6	18	17	15	40	–	–	Chipping
		9	100	2.6	18	35	27	200	218	0.08	
		9	100	2.6	18	52.5	40.4	–	–	–	
		9	100	2.6	18	70	54	170	–	–	
37	Basalt	72	50	3	45	14	9.3	190	–	–	Chipping
		72	50	3	45	24	16	235	444	0.13	
		72	50	3	45	34	22.6	275	250	0.07	
		72	50	3	45	34	27.3	330	182	0.05	
38	Concrete	1.7	50	2.5	150	10	8	200	–	–	
		1.7	50	2.5	150	25	20	250	300	0.15	
		1.7	50	2.5	150	40	32	440	167	0.08	
		1.7	50	2.5	150	50	40	550	91	0.04	

No.	Material										Notes
39	Concrete	1.7	50	2.5	165	10	8	125	—	—	
		1.7	50	2.5	165	25	20	160	429	0.21	
		1.7	50	2.5	165	40	32	270	137	0.08	
		1.7	50	2.5	165	50	40	365	118	0.06	
40	Concrete	1.8	50	2.5	150	10	8	120	—	—	
		1.8	50	2.5	150	25	20	170	300	0.15	
		1.8	50	2.5	150	40	32	360	158	0.08	
		1.8	50	2.5	150	50	40	430	142	0.07	
41	Concrete	1.8	50	2.5	150	10	8	110	—	—	
		1.8	50	2.5	150	25	20	160	300	0.15	
		1.8	50	2.5	150	40	32	275	130	0.06	
		1.8	50	2.5	150	50	40	345	143	0.07	
45	Concrete	4.6	50	2.4	160	3	2.5	95	—	—	
		4.6	50	2.4	160	10	8.3	115	700	0.35	
		4.6	50	2.4	160	25	20.8	175	250	0.12	
		4.6	50	2.4	160	40	33.3	265	167	0.08	
		4.6	50	2.4	160	47	39	260	—	—	
46	Concrete	3.5	50	2.5	165	3	2.4	158	—	—	
		3.5	50	2.5	165	10	8	180	637	0.32	
		3.5	50	2.5	165	25	20	235	273	0.13	
		3.5	50	2.5	165	40	32	345	167	0.08	
		3.5	50	2.5	165	47	37.6	405	117	0.06	Chipping
47	Concrete	3.5	50	2.5	160	3	2.4	154	—	—	
		3.5	50	2.5	160	10	8	175	667	0.33	
		3.5	50	2.5	160	25	20	310	370	0.18	
		3.5	50	2.5	160	40	32	225	178	0.09	
		3.5	50	2.5	160	47	37 6	205	350	0.17	
48	Concrete	3	50	2.5	155	3	2.4	105	—	—	
		3	50	2.5	155	10	8	125	700	0.35	
		3	50	2.5	155	25	20	165	375	0.18	Chipping

Contd.

Table 11 Contd.

(1)	(2)	(3)	(4)	(5)	(6)	(7)	(8)	(9)	(10)	(11)	(12)
49	Concrete	3	50	2.5	155	40	32	230	231	0.11	
		3	50	2.5	155	47	37.6	180	—	—	Chipping
50	Concrete	1.9	50	2.5	155	5	4	165	428	0.21	
		1.9	50	2.5	155	20	16	200	300	0.15	Chipping
		1.9	50	2.5	155	35	28	250	600	0.29	
		1.9	50	2.5	155	47	37.6	270	—	—	
51	Concrete	1.9	50	1.8	185	3	3.3	180	—	—	
		1.9	50	1.8	185	5	5.5	175	260	0.13	Chipping
		1.9	50	1.8	185	35	38.9	290	—	—	
		1.9	50	1.8	185	47	52.2	260	—	—	
52	Concrete	2.5	50	2.3	185	5	4.3	230	250	0.12	
		2.5	50	2.3	185	20	17.4	290	214	0.11	
		2.5	50	2.3	185	35	30.4	360	—	—	
54	Concrete	2.5	50	2.3	190	5	4.3	173	469	0.23	
		2.5	50	2.3	190	20	17.4	205	200	0.1	Chipping
		2.5	50	2.3	190	35	30.4	280	—	—	
		2.5	50	2.3	190	47	40.8	180	—	—	
55	Concrete	2.5	50	2.3	200	5	4.3	200	500	0.25	
		2.5	50	2.3	200	20	17.4	230	187	0.09	Chipping
		2.5	50	2.3	200	35	30.4	310	—	—	
		2.5	50	2.3	200	47	40.8	190	—	—	
56	Concrete	2.5	50	2.3	187	5	4.3	90	500	0.25	
		2.5	50	2.3	187	20	17.4	120	250	0.12	
		2.5	50	2.3	187	35	30.4	180	109	0.05	
		2.5	50	2.3	187	47	40.8	290	—	—	
57	Concrete	2.5	50	2.3	187	5	4.3	80	500	0.25	
		2.5	50	2.3	187	20	17.4	110	300	0.15	
		2.5	50	2.3	187	35	30.4	160	133	0.07	
		4.2	63	2.5	175	30	24	250	—	—	

No.	Material	(1)	(2)	(3)	(4)	(5)	(6)	(7)	(8)	(9)	Remarks
59	Concrete	4.2	63	2.5	175	45	32	160	280	0.14	
		4.2	63	2.5	175	60	48	370	162	0.08	
60	Concrete	2.7	60	2.6	168	2.5	1.9	120	—	—	
		2.7	60	2.6	168	12.5	9.6	160	500	0.25	Chipping
		2.7	60	2.6	168	25	19.2	110	—	—	Chipping
		2.7	60	2.6	168	37.5	28.8	90	—	—	Chipping
		2.7	60	2.6	168	50	38.4	150	208	0.1	
61	Concrete	2.7	60	2.6	165	2.5	1.9	70	—	—	
		2.7	60	2.6	165	12.5	9.6	85	667	0.33	Chipping
		2.7	60	2.6	165	25	19.2	140	227	0.11	
		2.7	60	2.6	165	37.5	28.8	230	139	0.07	
		2.7	60	2.6	165	50	38.4	150	—	—	
62	Concrete	2.7	60	2.6	165	2.5	1.9	70	—	—	
		2.7	60	2.6	165	12.5	9.6	85	666	0.33	Chipping
		2.7	60	2.6	165	25	19.2	110	500	0.25	
		2.7	60	2.6	165	37.5	28.8	160	250	0.12	
		2.7	60	2.6	165	50	38.4	100	—	—	
63	Concrete	2.3	110	2.5	135	12.5	10	45	—	—	
		2.3	110	2.5	135	25	20	90	278	0.14	Chipping
		2.3	110	2.5	135	37.5	30	150	208	0.1	
		2.3	110	2.5	135	50	40	195	277	0.14	
64	Concrete	2.3	60	2.5	185	0	0	40	—	—	
		2.3	60	2.5	185	12.5	10	60	625	0.31	
		2.3	60	2.5	185	25	20	105	278	0.14	
		2.3	60	2.5	185	37.5	30	160	228	0.11	
		2.3	60	2.5	185	50	40	225	192	0.09	
65	Concrete	2.3	60	2.5	185	12.5	10	60	—	—	
		2.3	60	2.5	185	25	20	100	312	0.15	
		2.3	60	2.5	185	37.5	30	150	250	0.12	
		2.3	60	2.5	185	50	40	210	208	0.1	
		2.3	60	2.5	185	0	0	45	—	—	
		2.3	60	2.5	185	12.5	10	70	500	0.25	

Contd.

Table 11 Contd.

(1)	(2)	(3)	(4)	(5)	(6)	(7)	(8)	(9)	(10)	(11)	(12)
66	Concrete	2.3	60	2.5	185	25	20	105	357	0.17	
		2.3	60	2.5	185	37.5	30	150	278	0.18	
		2.3	60	2.5	185	50	40	210	208	0.1	
67	Concrete	2.6	60	2.5	185	12.5	10	65	—	—	
		2.6	60	2.5	185	25	20	115	250	0.12	
		2.6	60	2.5	185	37.5	30	170	227	0.11	
		2.6	60	2.5	185	50	40	235	192	0.09	
68	Concrete	3.3	73	2.5	155	0	0	35	—	—	
		3.3	73	2.5	155	16	12.8	90	582	0.29	
		3.3	73	2.5	155	30	24	190	280	0.14	
		3.3	73	2.5	155	39.5	31.6	235	211	0.1	
		3.3	73	2.5	155	59	47.2	335	195	0.09	
69	Concrete	3.3	70	2.5	175	0	0	45	—	—	
		3.3	70	2.5	175	16	12.8	85	400	0.2	
		3.3	70	2.5	175	30.5	24	130	322	0.16	
		3.3	70	2.5	175	42.5	31.6	195	185	0.09	
		3.3	70	2.5	175	61	47.2	165	—	—	Chipping
70	Concrete	3.4	70	2.6	175	0	0	53	—	—	
		3.4	70	2.6	175	17	13	90	459	0.23	
		3.4	70	2.6	175	30.5	23.5	145	255	0.13	
		3.4	70	2.6	175	42	32.3	195	230	0.11	
		3.4	70	2.6	175	59.2	45.5	155	—	—	Chipping
		4	70	2.5	170	0	0	35	—	—	
		4	70	2.5	170	5.2	4.2	50	693	0.34	
		4	70	2.5	170	10.2	8.2	60	500	0.25	
		4	70	2.5	170	34.2	27.3	140	302	0.15	
		4	70	2.5	170	59.2	47.4	45	—	—	Chipping
74	Concrete	3.5	55	2.5	175	0	0	150	—	—	
		3.5	55	2.5	175	7	5.6	130	700	0.35	
		3.5	55	2.5	175	17.5	14	120	—	—	Chipping

48

No.	Material										
75	Concrete	3.5	55	2.5	175	27.5	22	115	—	—	Chipping
		3.5	55	2.5	175	37.5	30	150	286	0.14	
		3.5	54	2.5	175	0	0	100	—	—	
		3.5	54	2.5	175	7	5.6	110	700	0.35	
		3.5	54	2.5	175	18.5	14.8	135	460	0.23	
		3.5	54	2.5	175	28	22.4	120	—	—	
		3.5	54	2.5	175	39	31.2	155	314	0.16	Chipping
76	Tuff	9	40	3.2	90	5	3.1	35	—	—	
		9	40	3.2	90	15	9.4	55	500	0.2	
		9	40	3.2	90	25	15.6	90	286	0.11	
		9	40	3.2	90	35	21.9	135	222	0.09	
77	Tuff	9	40	3.2	90	5	3.1	35	—	—	
		9	40	3.2	90	15	9.4	55	500	0.2	
		9	40	3.2	90	25	15.6	75	500	0.2	
		9	40	3.2	90	35	21.9	100	400	0.16	
78	Tuff	9	80	3.2	90	5	3.1	60	—	—	
		9	80	3.2	90	15	9.4	75	667	0.26	
		9	80	3.2	90	55	34.4	235	250	0.1	
		9	80	3.2	90	75	46.9	340	190	0.07	
79	Tuff	9	80	3.2	90	5	3.1	39	—	—	
		9	80	3.2	90	15	9.4	57	555	0.22	
		9	80	3.2	90	45	28.1	370	—	—	
		9	80	3.2	90	75	46.9	440	—	—	
80	Secondary quartzite	135	40	3.2	60	0	0	40	—	—	Chipping
		135	40	3.2	60	10	6.2	55	667	0.1	Chipping
		135	40	3.2	60	20	12.5	100	444	0.09	

50

type closely describes such a relationship

$$v_T = 750 \exp\left(-0.04\frac{r}{r_0}\right). \tag{2.2}$$

The results obtained show that the chosen model of elastoplastic medium is applicable for solving problems encountered in blasting mechanics.

The laboratory experiments give specific quantitative indices, apart from the qualitative picture of breakage by blasting. The exact value of growth rates of cracks is essential for optimising many of the technological parameters. Thus for breakage of ore with presplitting of a narrow slope area, it is very important to know the widening rate of crack surfaces (screening cavity) as well as the rate of crack movement.

Experimental results indicate that during breakage of ore (rock), the failure front moves from the cavern towards the compensating cavity, i.e., along LLR. In this direction itself a fall in the intensity of crack growth is observed. If at a relative distance of 4.2–5.6, the rate of crack movement is 690–700 m/s, then at $r/r_0 = 32.0$–33.3 it equals 158–231 m/s (see Table 11).

In breakage of basalt with a strength of 7.2 on the Protod'yakonov scale, the intensity of rate of crack movement falls approximately by 11 times, i.e., 444–40 m/s, in the area confined within the relative distance 16–84.6. In this case, the widening rate of crack surfaces is equal to 10–82 m/s, for a change in relative distance between 37.6–2.4.

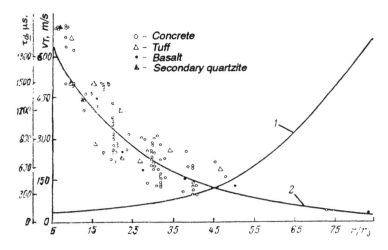

Fig. 16. Dependence of change in (1) durability of the body and (2) rate of crack movement on the relative distance

The ratio of rate of crack movement to the velocity of longitudinal wave in the media tested was found to be 0.03–0.35 as the relative distance varied from 73.0–4.2. The means of these rates tend largely to 0.07 of the velocity of longitudinal elastic waves.

The principal achievement in the experimental investigations discussed above lies in the fact that the established rates of crack movement as well as the rate of movement of massif are useful in engineering calculations of parameters of breakage by blasting.

2.2 Branching of Cracks

The aspect of branching in cracks caught the attention of researchers several years ago. The initial theoretical and experimental studies of branching in cracks are found in many publications for example those of E. Ioffe, J. Irwin and J. Shardin. Polymers were used as the object of their investigations.

The branching of cracks in polymers and metals under dynamic (explosive) loading has been studied in detail, for example by V.M. Finkel. Still, an exact picture of the development of cracks in rocks does not emerge despite the many publications available. This can be ascribed to the difficulties in conducting a dynamic experiment to study the branching of cracks in optically inactive media.

Very often, evaluation of the processes of initiation, growth and branching of cracks in rocks is based on experimental results of breakage in models made from acrylic plastic. This, obviously, is not strictly correct. Actually, during a blast in an acrylic plastic model, an area deformed by the effect of the blast wave is clearly demarcated around the cavern. The transparency of the acrylic plastic is affected in this zone and therefore a clear picture cannot be obtained on the branching of cracks. The concentration of deformations at surfaces often gives an illusive picture about the formation of macrocracks.

It is very essential to obtain a picture of branching of cracks under laboratory conditions in order to study the effect of blasting of explosive charges on the quality of ore (rock) fragmentation. Moreover, this allows reduction in volume of laborious experimentation under field conditions.

The practice of breakage by blasting shows that the granulometric composition of ore (rock) is represented by irregularly shaped fragments. Fragments of similar form but smaller in size by the quantity of chosen geometric scale of modelling are obtained as a result of blasting in blocks made from equivalent materials, in which almost no cracks occurred. This indicates the hierarchy of fractures in the disintegration process of rock (ore). V.N. Rodionov observed that in a blast, the crushing zone of thiosulphate is also pierced by several chaotically directed microcracks.

A method for recording the growth of cracks during the breakage of rocks by blasting was developed in 1978 [3]. This method enabled photographing the initiation, growth and branching of cracks at the surface of test objects. It has already been shown above that during blast breakage the velocity of the crack growth is several times lower than the velocity of the blast wave. Consequently, macrocracks develop during the period of piston effect of gases. Macrocracks do not develop simultaneously on the surface of the blocks, however; major cracks appear first followed by branched ones. Pulverisation of one of the blocks of cemented fill begins with the appearance of three leading cracks. Two of them are approximately oriented along the direction of charge and their ends diverge. The third crack intersects these two roughly at 90°. In the fourth frame, from the end of the left crack, initially a branched crack takes off, which propagates in the opposite direction. As a result of some re-dislocation, five cracks remain in the sixth frame, which are oriented in a direction leading from the centre to the boundary of the block. The mutual intersection of cracks, with different widening rates of surfaces, is seen in the ninth frame. In later frames, the distances between crack surfaces grow very fast, especially in the section opposite the charge. In the curved portions, cracks appear branched. The widening rate of surfaces of a branched crack was initially found to be 3 m/s but later increased up to 6 m/s.

The branching of cracks during blasting of charges of varied configuration is of great significance in research. Thus, in a point (concentrated) charge, major cracks originate at the surface of a sphere in the form of a 'rose'. In an elongated charge, the major cracks are oriented along its directrix, while branched cracks are situated at various angles to the axis. The multifaceted schemes of branching in cracks indicate that breakage (crushing) depends on material properties as well as charge configuration.

In all the experiments, the rate of movement of major cracks was found to be more than that in branched cracks (see Table 9). At each branching, the crack became diverted, maintaining its initial direction, however. The rate of movement of a branched crack fell initially but later increased and became equal to the velocity of the major crack.

On comparing the types of branching in cracks, certain similarities are readily discernible. In spite of large differences in physicomechanical properties, the tested materials (basalt, marble, tuff, concrete etc.) exhibited relatively similar types of crack branching.

The change in velocity of a crack, i.e., the fact of existence of acceleration, indicates the presence of forces of inertia associated with them, which obviously influence the process of branching.

It is common knowledge that breakage by blasting is a controllable process. In this connection, the formulation of experiments with air-deck charges is of interest. Despite much research experience, there is no unanimity of concept even today on the effect of charge configuration on the breakage (crushing) of the massif.

Blocks of red tuff, which contained no cracks, were blasted. Here, all the similar parameters except the charge configuration were taken as constant (depth of hole 80 mm, diameter 2 mm, weight of TEN explosive 124 mg, LLR 25 mm). In the first block, a continuous charge of 40-mm length was placed, while in the second block, the charge was separated into two parts, 28 and 12 mm long respectively, by an air deck (10 mm).

The process was photographed with an exposure time of 20 µs and interval between snaps of 100 µs. The leading cracks were seen in the stage of formation in the third frame, i.e., when the blast effect duration was about 260 µs.

The rate of crack movement in the first and third blocks was equal to 96 m/s. However, in subsequent frames, the failure pattern of the blocks was entirely different. When a continuous charge was used, a major crack appeared (frame 3) at the opposite side of the charge column. At the branching off points, a deviation of this crack was noticeable. A major crack also appeared when an air-deck charge was used. However, at the lower part of the bench, cracks were represented by a 'rose' that covered almost two-thirds of the body section. The results of experiments in this series showed that the use of air-deck charges the degree of crushing of the massif and also indicated the combination of types of characteristic branching in cracks for spherical and drillhole charges.

In rock breakage by blasting, the quality of fragmentation depends on the chosen parameters, i.e., blasting method, type of explosive and charge configuration. The last factor has considerable influence since it ensures an improved degree (by 1.2–1.3 times) of fragmentation. Charges with axial cavities have found wide use in underground metal mines, for example in Krivoy Rog.

2.3 Delays Occurring in Failure of the Massif

Breakage by blasting involves control of ore fragmentation, taking into account the requirements imposed on the stability of structural elements of the room (block) under the effect of loading. In such conditions, the knowledge base of the strength behaviour of a body has to be enlarged, especially consideration of time as a characteristic of the processes of deformation of the medium and its failure. The time-dependent strength properties of rocks and concrete can be taken into account by using data from experimental investigations of the failure process of the massif.

Experimental data have shown that the massif gets crushed at a relatively later stage in the process, after undergoing significant deformation. Therefore, no attempt to describe breakage based only on the data of stresses σ in the massif, for example, using the time-based criterion of strength proposed by Zhurkov [19] or wave-time criterion of the type,

$$\int \sigma^n dt = \text{const, } (n = 1 \text{ or } 2)$$

has been successful. Further, the equation of kinetics of crack growth $v_T = v_T(\sigma)$ also cannot be used with approximations in a massif, in which they happen to be negligibly small, while σ corresponds to the undisturbed massif. Based on this, it is considered that the delay in failure is related to the variation in structure of the massif over the time period of blast effect. Durability (longevity) of the body τ_d is a time interval that demarcates the moment of initiation of crack from the moment of application of load; it is evaluated by the time lapsed until the initiation of the major crack, i.e., until the material ruptures.

The durability of a body is proportional to the reciprocal value of the rate of crack growth. If the interrelation between these parameters is established, then by knowing the rate of movement of a crack, it is possible to compute the durability of a body.

As an example, let us consider the effect of rate of movement of crack on the durability of materials tested. The relationship between v_T and relative distance can be approximated by (2.2). The established relationship between durability of the material and relative distance is expressed by

$$\tau_d = 197.3 - 10.8(r/r_0) + 0.4(r/r_0)^2. \qquad (2.3)$$

The simultaneous solution of (2.2) and (2.3) shows that the relationship between durability of tested materials and rate of movement of cracks is approximated by the expression

$$\tau_d = 267.3 + 4.1 \ln v_T - 0.4 \ln^2 v_T. \qquad (2.4)$$

It follows from this expression that with an increase in v_T the curve approaches the x-axis asymptotically.

The accumulated experimental data (see Section 2.1) give a quantitative assessment of the rock breakage process. The breakage process itself can be represented as development and branching of dynamic cracks under the blast effect when a pressurised gas is released into the cavity causing deformation and disintegration of the massif.

The experimental data also show that at a relative distance of 4.2, the rate of crack movement attains the ultimate value of 750 m/s; however, at $r/r_0 = 60$, the rate is considerably lower (60 m/s). This indicates that in the breakage of rocks by blasting commercial explosives the front of disturbances advances much farther than the front of movement of directed (major) and branched cracks.

During the experimental studies, weak spots (holes) were created in the massif being broken. In the process, the sequence of breakage of graphitic rods was disturbed, i.e., the major cracks started to develop at the weak spots earlier. This in no way contradicts the established patterns of movement of the failure front from the charge towards the exposed surface. The advance failure occurred locally in those areas of the massif which were weakened or overloaded.

2.4 Problem of Strength/Deformation Properties of a Solid Massif

The conventional design techniques for methods of working with back-fill do not take into account all the mechanical properties of cemented fill mixtures, which determine the mutual influence of blast and backfill, back-fill and rock massif, support and backfill. Only the strength index of rock cubes under uniaxial compression are taken into account in computations. If other significant properties of cemented fill mixtures are not taken into cognizance, this results in overrated design norms. If the norm happens to be excessive, the filling cost increases; conversely, if the norm is inadequate, the stability of the artificially filled massif is endangered. For normal geotechnical conditions, the measure suggested above is justified by current mining practice, which provides input data for the experimental determination of design norms. As mining depth increases, and with the adoption of breakage by blasting, a non-standard situation emerges, which complicates the design. This problem can be solved by taking into account all factors that determine the combined working of blast and massif at the rear of the contour (fill), including the indices of resistance and deformability of the fill material.

A large volume of research experience, with results tested in metal mines, is presently available. Effective methods for controlling the blast effect have been developed, which are directed towards changing the granulometric composition of ore by taking into account the requirements of subsequent treatment [5, 25]. But the phenomena preceding failure in macroscopic volumes have not been adequately studied. Let us now examine this aspect of the breakage process by blasting, which is associated with deformation/strength properties of polycrystalline materials, such as rocks, concrete (fill) etc.

One method that simultaneously monitors the stresses and deformations during a blast is based on experiments using special gauges. These experiments involved fixing the gauge in the model block (Fig. 17), blasting a charge and recording the stresses developed during propagation of the blast wave. In so doing, the necessary and sufficient conditions of similarity criteria (1.1), (1.4), (1.8) and (1.19) were satisfied. A compensating cavity was not created in the model block and the lateral displacement of particles was constrained by an adequately reliable system (metallic walls of 15-mm thickness).

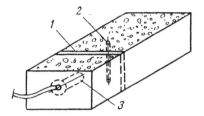

Fig. 17. Experimental set-up for studying the effect of crack on stresses.

1—crack; 2—charge; 3—gauge.

In its initial state, the rock massif is disturbed by cracks of various sizes. The maximum stress at the wavefront happens to be a parametrically assigned function of two variables—strength of the medium R_{st} and crack width, b, i.e.,

$$\sigma_{max} = f(R_{st}, b).$$

Let us initially establish the effect of strength of the medium on the maximum stress during a blast at a known relative distance, for example, 40. Results of all the experiments on models are given in Table 12 and are described by the empirical relationship

$$\sigma_{max} = 51.16 + 3.87R_{st} - 0.03R_{st}^2, \tag{2.5}$$

which has a correlation coefficient of 0.9. The relationship (2.5) is graphically represented in Fig. 18.

Let us discuss the experimental results obtained from the point of view of modern concepts of breakage of rocks and concrete. Thereafter we shall highlight their usefulness in solving problems of metal mining technology.

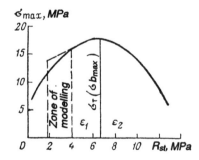

Fig. 18. Stress versus strength of material.

$\sigma_{b\,max}$—maximum stress corresponding to the boundary of formation zone of microfailures

Table 12

σ_{max}, MPa	R_{st}, MPa	σ_{max}/R_{st}	σ_{max}, MPa	R_{st}, MPa	σ_{max}/R_{st}
8.8	1	8.8	16 8	8	2.1
11.7	2	5 85	15.6	9	1.72
14	3	4.67	13.9	10	1.39
15.8	4	3.95	11.4	11	1.04
17	5	3.4	8 5	12	0.71
17.6	6	2.93	4	13	0.31
17.5	7	2.5			

During a blast in blocks of a model, when lateral movement of particles is constrained, stresses increase with an increase in strength of material up to 5 MPa. In Fig. 18, in the segment of deformation beginning at strength of material 5 MPa and reaching up to 6.7 MPa, the growth of stresses is moderate; with a subsequent increase in strength ($R_{st} > 6.7$ MPa), the stresses decrease markedly. From this, the relationship between deformations and stresses in the ascending and descending branches of curve $\sigma_{max} - R_{st}$ can be inferred. If average deformations are related to the moment of attaining maximum stress σ_{max}, then the following is observed. For strains ε_1, corresponding to the left branch, an increase in σ_{max} is noticed with enhanced strength of equivalent material. At strains ε_2, corresponding to the right branch, σ_{max} decreases with an increase in R_{st}. The value σ_T characterises the boundary of disturbance of continuity in the material due to accumulation and growth of microcracks. Concentration of initial minicracks leads to subsequent damage of the structure. The merging of microscopic cracks results in the occurrence of visual cracks and later leads to rapid disintegration. The strength condition is determined by the volume of accumulated microfailures. The transition from accumulation of microcracks and small clusterings to sharp enlargement is associated with satisfying the concentration criterion [19].

The variation in relative intensity of the blast wave with increase in static strength of material is shown in Table 12.

The relative intensity of blast-induced stresses in equivalent material fell 8.8 times with a change in R_{st} from 1 to 13 MPa (Fig. 19).

With an increase in failure in the microscopic volumes, the energy of a blast wave increasingly dissipates, which results in reduced stresses in the material. The strength σ_T, corresponding to the boundary of the formation zone of microcracks, is determined by a fall in the value of the maximum stress. σ_T can be found out from

$$\sigma_T = \sigma_{b\,max} - \sigma_{max},$$

where $\sigma_{b\,max}$ — maximum stress, corresponding to the boundary of formation zone of microfailures at a relative distance of 40;

σ_{max} — maximum stress, corresponding to the given strength of material R_{st} at the same relative distance.

The computed values of σ_T are given below.

R_{st}, MPa	7	8	9	10	11	12	13
σ_T, MPa	0.1	1.8	3	4.7	7.2	10.1	14.1
σ_T/R_{st}	0.01	0.22	0.33	0.47	0.65	0.84	1.08

The boundary of the formation zone of microfailures (Fig. 20) was established on the basis of the given experiments by calculating the ratio σ_T/R_{st}. According to the defined boundary of the formation zone of microfailures, the value of the ratio σ_T/R_{st} was proportional to the increase in material strength.

From the results obtained, it follows that the stress σ_T, corresponding to the boundary of the formation zone of microfailures, is an important characteristic of the material. The constrained movement of the massif around the contour impedes any increase in its volume from the moment of attaining σ_T and therefore the strength of particles increases.

An analysis of the measured stresses with an increase in the strength of material enables one to demarcate a zone of normative strength of the fill massif that characterises the upper boundary of pressure at 6.7 MPa. If the strength of the fill massif exceeds 6.7 MPa, the maximum stress will attenuate due to an increase in deformation. With an increase in R_{st} from 0.6 to 6.7 MPa, an increase in σ_{max} from 6 to 20 MPa was observed, while the time of wave attenuation remained at the level of 200 µs. In the ε_2 zone, where attenuation of σ_{max} ranged from 20 to 6 MPa, the time of wave attenuation was 200-80 µs.

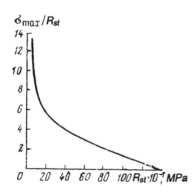

Fig. 19. Relationship between relative intensity of blast wave and static strength of material

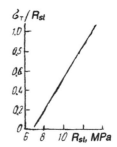

Fig. 20. Relationship between σ_T/R_{st} and R_{st}

Considering the constraint equation $\sigma_{max} - R_{st}$ in the interval $R_{st} =$ 1.7–4 MPa (see Fig. 18), we obtained the equation for a step-wise linear function,

$$\sigma_{max} = 117.04 + 1.03R_{st}.$$

From this it follows that at a modelling scale of 1/50–1/30 for breakage by blasting, the strength of material (used in the model) under uniform compression corresponds to 1.7–4 MPa. With an increase in R_{st} from 1.7 to 4 MPa, there is a corresponding increase in σ_{max} from 13.4 to 15.8 MPa. Here it becomes possible to eliminate the scale effect since modelling is done by the method of equivalent materials. In certain experiments, when the strength of material exceeded 6.5 MPa, breakage of the medium did not occur due to the fact that the energy of the explosive charge was commensurate with the internal energy of the material.

Let us consider the effect of strength of material on energy losses in the blast wave, based on the calculation methodology suggested by G.I. Pokrovskii.

The experimental and computed values of relative energy remaining in the blast wave in the model at different strengths of equivalent material are given in Table 13. The relationship is expressed by a parabola of the type

$$U/U_0 = 0.0916R_{st} - 0.0007R_{st}^2 - 0.257.$$

From this empirical relationship, it may be inferred that the energy loss in a blast increases with variation in the strength of the material. When the strength of the material is 6.6–7.0 MPa (Fig. 21), the relative fraction of energy remaining in the blast wave is comparatively higher (2.9%) and in some experiments was even 5.3%.

To study the influence of the extent of discontinuity in a massif on the intensity of blast wave, cracks were initiated in blocks of the model. Quartz sand was used as a filler in the cracks. Gauges were installed in

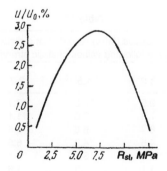

Fig. 21. Relative energy U/U_0 remaining in the blast wave versus strength of material.

Table 13

R_{st}, MPa	σ_{max}, MPa	r/r_0	r_p	U/U_0, %
0.6	8.1	40	240	0.46
0.6	8.2	40	240	0.46
0.9	12.4	41	160	0.49
0.9	5.4	40	350	0.15
1	7.6	39	220	0.54
1	7.4	40	250	0.41
1.5	10.2	37	190	0.74
1.8	10.2	39	190	0 86
1.8	13.2	37	170	1.03
2.4	19	40	105	1.03
2.6	13 3	40	165	1.42
3 4	14.8	40	147	2.01
4.6	17 8	40	125	3.27
4 6	11	40	180	1.06
5 2	15.9	40	140	2.33
6.6	20.2	40	100	5.3
7	18.5	40	118	3.93
7.5	18	39	120	3.27
8.5	18	41	120	3 98
9	15.3	43	150	2.35
9.5	13.5	40	170	1.3
9.5	13.6	40	170	1.3
10	10	44	190	1.24
10 6	11.2	40	180	1.09
11	10 2	40	190	0.93
12 7	8.7	39	210	0.64
12.7	8 6	39	210	0.64

the blocks at a relative distance of 40 mm and the diameter of the charge (TEN) was 1.8–2 mm. The mean values of maximum stresses, based on the results of 3 to 7 experiments, are given in Table 14.

An evaluation of the effect of crack width on σ_{max} for a known strength of material yielded a family of divergent straight lines as the crack width

Table 14

Crack width, mm	Stress (MPa) recorded at a distance $r/r_0 = 40$, for the following values of static strength of material, MPa					
	0.7	1 2	1.5	2.7	5 0	6.5
0	7.6	9.5	10.3	13.1	17	17.6
2.5	6 2	7.8	8.6	11.1	14	14.5
5	5	6	6.8	8.6	12.2	11.9
7 5	3 7	4.5	4.9	6.8	8.1	8.5
10	2	2.6	3.2	4	4 9	5.8
12.5	0.8	1.1	1.3	1.6	2.3	2.6

b reduced (Fig. 22).

Clearly, the relationship between σ_{max} (maximum stress) and *b* (crack width) and R_{st} (strength of material) is of the form

$$\sigma_{max} = m \pm nb. \qquad (2.6)$$

A functional relationship exists between R_{st} and the coefficients *m* and *n* (Fig. 23), which is described by equations of the type:

$$\left.\begin{array}{l} m = AR_{st} + BR_{st}^2, \\ n = CR_{st} + DR_{st}^2 \end{array}\right\}. \qquad (2.7)$$

The unknown parameters in these equations are determined by the method of multiple correlation:

$$m = 7.48R_{st} - 0.0816R_{st}^2, \qquad (2.8a)$$

$$n = 0.512R_{st} - 0.056R_{st}^2. \qquad (2.8b)$$

Taking into account (2.6), (2.8a) and (2.8b), the theoretical line of regression is of the form

$$\sigma_{max} = 7.5R_{st} - 0.08R_{st}^2 - (0.05R_{st} - 0.05R_{st}^2)b. \qquad (2.9)$$

This expression establishes the relationship between the strength of material ($0.6 \leqslant R_{st} \leqslant 13$ MPa) and the extent of widening of surfaces of a crack (with a weak filler) as well as the maximum stress that develops in the massif when the blast wave passes through at a relative distance of 40. It needs to be borne in mind that this is only relevant to metal mining practice that employs cemented backfill.

Reliability of the method of working with backfill is characterised by stability (failure-free nature) of its constructive elements during a blast

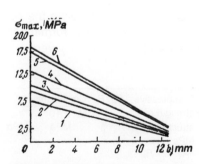

Fig. 22. Maximum stress σ_{max} versus crack width *b* for strength of material R_{st}. (1) 0.7; (2) 1.2; (3) 1.5; (4) 2.7; (5) 5.0 and (6) 6.5 MPa.

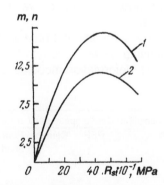

Fig. 23. Relationship between coefficients (1) *m* and (2) *n* and strength of material.

and the attenuation of the deformation processes. Much experience has now been gained in the application of cemented backfill in metal mines. With considerable increase in strength of material (more than 7 MPa), the volume of breakage of the artificial massif increases and therefore dilution is not significantly affected.

Specialists have paid attention in recent years to the deformation properties of mixtures used as fillers in the stoped-out area in metal mines. The practical significance of results of such investigations is determined by the requirements imposed on the flexibility of technological schemes for stoping in complex geomining conditions, and in particular in unstable ores and host rocks [24, 26].

Deformation properties of mixtures were studied using prisms of 7 × 7 × 28 cm. The strength of the material under compression or tension was determined by testing cubic specimens with a 7-cm edge and cubes of the same size with sloped edges. The metallic mould was filled with the mixture of chosen composition (Table 15) having a swelling space of 12.3–12.5 cm. The specimens were removed from the mould the next day and transferred to a chamber for storage in wet sawdust at a temperature of $20 \pm 2°C$. After 28 days, the specimens were tested in a 100-T Universal Testing Machine with a rate of loading of 0.3 MPa/s. The longitudinal deformation of each prism under compression was recorded using wire strain gauges of 20-mm base, attached at the two opposite edges of the specimen. During testing, the prism was loaded at increments of $0.1R_{st}, 0.2R_{st} \dots, 0.5R_{st}$ with a pause of 3 min at each stage. The ultimate strength of the prism was taken as 0.8 of the cube strength, i.e., $R_{st} = 0.8R_K$.

Deformations were recorded at the beginning and end of each stage of loading. At a load of $0.8R_{st}$, the specimen was relieved for 5 min and the residual deformation was recorded; it was later loaded repeatedly until failure.

The experimental data were processed and curves showing the relationship between compressive stresses and longitudinal deformations $(\sigma - \varepsilon)$, deformation modulus and compressive stresses $(E - \sigma)$ were ob-

Table 15

Composition of fill	Consumption per 1 m³ of fill, kg				Compressive strength of structure, MPa $\times 10^{-1}$		R_{st}/R_K	Coefficients	
	PEF	Cement	Pumice sand	Water	R_K	R_{st}		$M_y \times 10^{-5}$, MPa	Y_s, MPa
1	—	200	1050	380	20.3	17.4	0.86	0.198	22.5
2	400	—	850	450	24 6	20.5	0.83	0.222	26.6
3	300–400	70	800–900	450	37.3	31.3	0.84	0.295	40.7

tained. These are analogous to the investigations of A.A. Arakelyan. The relationship between deformation modulus and compressive stress can be expressed with reasonable accuracy by:

$$E = M_y \left(1 - \frac{\sigma}{Y_s}\right),$$ (2.10)

where M_y and Y_s are parameters of deformation modulus obtained from the results of experimental data. M_y is the conditional initial elasticity modulus characteristic of the given material in an unstressed state. Y_s is the conditional yield strength of the material, which is more than its prismatic strength (R_{st}). Integrating the experimental curves of deformation moduli, we obtain an expression for relative deformations ($d\varepsilon = d\sigma/E$)

$$\varepsilon = \int \frac{d\sigma}{M_y \left(1 - \frac{\sigma}{Y_s}\right)} = -\frac{Y_s}{M_y} \ln \left(1 - \frac{\sigma}{Y_s}\right),$$ (2.11)

which is compared with the experimental data. (If the numerator of the function under integral happens to be the derivative of the denominator, usually the integral is equal to the logarithm of absolute quantity.)

The relationships between relative deformation and stress σ_{st} (Fig. 24a) and also between modulus of deformation and stress (Fig. 24b) for cemented fills of compositions 1, 2 and 3 (Table 16) are shown graphically.

The numerical values of M_y and Y_s contained in (2.10) are computed using the method of mathematical statistics.

The relationship between M_y and Y_s versus compressive strength of the fill structure is expressed by the formulae:

$$M_y = \frac{10^5 R_{st}}{64.5 + 1.35 R_{st}};$$ (2.12)

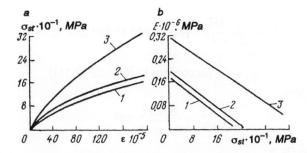

Fig. 24. Deformation properties of mixtures based on lithographic pumice
1 to 3—composition numbers of fill.

Table 16

Composition of fill	Compressive strength of structure, MPa $\times 10^{-1}$	Initial modulus of deformation, MPa $\times 10^{-1}$	Deformation modulus (MPa $\times 10^{-1}$) as a tangent of angle of slope			
			Tangential to line of deformation, at		Secant to line of deformation, at	
			$E_{0\,2R_k}$	$E_{0\,5R_k}$	$E_{0\,2R_k}$	$E_{0\,5R_k}$
1	20.3	20,000	16,700	12,300	18,000	15,500
2	24.6	22,500	19,300	14,000	21,300	17,700
3	37.3	30,000	25,200	18,200	28,000	23,200

$$Y_s = 1.35 R_{st}. \tag{2.13}$$

At known values of M_y and Y_s and taking $R_{st}/R_K = 0.85$, we determine the modulus of deformation E and longitudinal deformation ε according to Table 16. At $R_{st} = 0.85 R_K$ and $\sigma_{st} = 0.85 \sigma_K$, we obtain

$$E = \frac{0.74 R_K}{56 + R_K} \left(1 - \frac{\sigma_K}{1.3 R_K}\right) 10^5, \tag{2.14}$$

$$\varepsilon = -(84 + 1.5 R_{st}) \ln \left(1 - \frac{\sigma_K}{1.3 R_K}\right) 10^{-5}. \tag{2.15}$$

Compressibility of the structure (ultimate compressibility) at $\sigma_K = R_K$ is determined by the formula

$$\varepsilon_{max} = 1.47(84 + 1.5 R_K) 10^{-5}. \tag{2.16}$$

Stress in backfilled structures is usually found to be (0.2–0.5) R_K. Let us determine the deformation modulus of stresses $\sigma_K = 0.2 R_K$ and $\sigma_K = 0.5 R_K$, as tangent and secant to the line of longitudinal deformations $(\varepsilon - \sigma_K)$.

Tangent of modulus of deformation

at $\sigma_K = 0$,

$$E_0 = \frac{0.74 R_K}{56 + R_K}, \text{ [according to formula (2.14)]}$$

$$\varepsilon_0 = 0; \text{ [according to formula (2.15)]}$$

at $\sigma_K = 0.2 R_K$

$$E_{0.2R_K} = \frac{0.63 R_K}{56 + R_K} 10^5; \tag{2.17}$$

at $\sigma_K = 0.5 R_K$

$$E_{0.5R_K} = \frac{0.455 R_K}{56 + R_K} 10^5. \tag{2.18}$$

Secant deformation modulus:

at $\sigma_K = 0$

$$E_0 = \frac{0.74R_K}{56 + R_K} \cdot 10^5, \; \varepsilon_0 = 0;$$

at $\sigma_K = 0.2R_K$

$$E_{0.2R_K} = \frac{0.7R_K}{56 + R_K} \cdot 10^5; \tag{2.19}$$

at $\sigma_K = 0.5R_K$

$$E_{0.5R_K} = \frac{0.5R_K}{50 + R_K} \cdot 10^5. \tag{2.20}$$

The deformation moduli of tested materials were computed according to (2.14)–(2.20) and the results are presented in Table 16.

The data from experimental investigations show that the deformation modulus of a complex composition is comparatively higher. For example, the deformation modulus of fill containing 70 kg/m^3 of cement with 300–400 kg PEF as additive (composition 3) is 1.5 times higher than the deformation modulus of fill containing 200 kg of cement (composition 1). The yield strength of the third composition is twice as great as that of the first composition for the same setting time; this happens to be an impediment for the development of cracks during exposure of an artificial structure by the drill-blast method.

A light cemented fill based on (lithographic) pumice stone was used for the first time in a metal mine of Armenia [2]; this enabled the precious ores from different levels to be mined without hindrance and almost without losses. The mine is being worked under complex geomining conditions, using the cut-and-fill method with descending slices and free-steered vehicles. This method was perfected and introduced in the mine in 1973–1978. The output per manshift (OMS) rose from 0.9 to 15 m^3/manshift on changing over from the ascending method to the descending method and the possibility of further increase in OMS is not unlikely.

When the artificial structure based on composition 1 (Table 15) is exposed from a horizontal working of 16–20 m^2 cross-sectional area located below, the working is supported by closely spaced wooden frames at 1.5–2 m. Experience has shown that the use of composition 3 (see Table 15) allows an increase in the span of exposure of the filled structure up to 8 m and the cross-section of the stoping face can be enlarged to 30–40 m^2.

This becomes possible due to the glass-fibrous nature of the material, which enhances the deformability of the filled massif [26]. The normative strength of the massif should not exceed 6.7 MPa.

When the load on the filled structure exceeds its ultimate strength, consideration of the residual strength and deformability of a partially failed structure, loosened by cracks formed and developed due to quasiplastic deformations, assumes special significance. The established residual strength of tested materials is given in Tables 17 and 18.

The loading of backfill by known and gradually increasing values of

Table 17

Components of filling material	Consumption of components per 1 m³ of fill mixture, kg	Structural density, g/cm³	Ultimate strength, MPa	Residual strength MPa	%	Age, days
Mill tailings from Kafansk plant	740	1 8	1 8	0.73	41 3	60
			2.14	0.78	36 5	60
PEF	520		1.91	0.79	41 0	90
Water	540		2.06	0.76	36.9	120
			2 31	0.80	34.6	120
Mill tailings from Kadzharansk plant	790	1.84	3.54	1.26	35 6	60
			3.98	1 28	32 2	90
PEF	550		3.52	0.89	25 3	120
Water	500		3.95	1.08	27.3	120
Pumice sand	850	1 7	5 41	1.23	22 7	28
PEF	400		4.85	1 39	28.7	28
Water	450		5 61	1.68	29.9	30
			5 77	2.05	35.5	30
			6.31	1.57	24 9	60
			6 83	1.57	23.0	60
			8.41	2.46	29.3	90
			7.96	1.88	23 6	90
			7.87	1.2	15.2	90
			7 27	1.5	20.6	180
			5.30	1 92	36 2	180
Quartz sand	1390	2.3	6.43	0.51	7 9	30
Cement	470		7.55	0.5	6.6	35
Water	440		7.41	0 97	13.1	90
Mill tailings of Kafansk plant	350	1.7	1.05	0 35	33 3	28
			1.06	0.38	35.8	28
Mill tailings of Kadzharansk plant	350		1.06	0 43	40 5	28
				0.74		
PEF	460		1.40		52.8	60
Water	540		1.36	0.65	47 7	60

compressive deformations is characterised externally by the following stages: increased resistance of material; retardation in growth of resistance as it approaches the ultimate strength value right up to the constant value of resistance (plasticity zone) when the assigned deformation is increased; rapid fall of resistance until the value of residual strength of material; subsequent slow reduction (sometimes increase) of residual strength in a wide range of assigned deformation.

The deformation curves up to and beyond the ultimate strength of materials are shown in Fig. 25 (a, b, c—filled structure, d—tuff). The specimens were tested in a UMM-5 Universal testing machine.

The deformation of specimens of fill material exhibited a simultaneous linear development of elastic and residual deformations when they were loaded up to their ultimate strength. When loading reached the ultimate

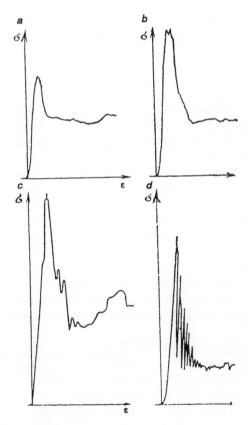

Fig. 25. Curves showing deformation of (a, b, c) structure and (d) tuff when loaded beyond ultimate strength

strength value, a zone of plastic deformation was formed with development of minor cracks and residual deformations exceeded 2–3 times the extent of deformations observed before reaching the ultimate strength. The fall in load-bearing capacity beyond the zone of plasticity was completed by the stabilisation of residual strength at (a) 41.0, (b) 32.2 and (c) 36.2%.

The Artiksk tuff deforms almost elastically when subjected to loading up to ultimate strength (7.74 MPa). Once the ultimate strength is reached, many major cracks suddenly develop in the specimens and their resistance falls sharply to 1.48 MPa.

The use of cemented fill of high strength characteristics undoubtedly enhances the effectiveness of the method of working with backfilling of a stoped-out area. In such a case, breakage by blasting large-diameter holes is possible, without leaving layers of ore at the contact.

2.5 Problem of Fractional Participation of Blast Wave and Piston Effect of Gases in the Breakage Process

Today many specialists are paying attention to the partial participation of the blast wave and the piston effect of gases in the breakage of rocks. As it becomes increasingly more important to enhance the quality and total recovery of ores from the earth's crust, to provide effective technology the assessment of partial participation of the blast wave and gases acquires greater significance. Actually, these are the factors which largely determine the volume and nature of breakage. This is an important stage in explaining the breakage mechanism of rocks by blasting, for finding the means to enhance the efficiency of currently used explosives and developing more effective explosives for the mining industry.

It is well known that a rapid transformation of the explosive in a narrow cavity leads to the initiation of a shock wave which attenuates over distance from the blast site and converts into a stress wave. The cavity expands due to the piston effect of highly compressed gases either immediately after initiation of the shock wave or simultaneous with it. It should be borne in mind that in the breakage of rocks by blasting charges of chemical explosives, no shock wavefront occurs. Hence the shock wave

Table 18

Rock	Structural density, g/cm^3	Ultimate strength, MPa	Residual strength	
			MPa	%
Grey tuff	1 5	16.1	4 18	25.9
Black tuff	1.6	35.5	6.77	19.1
Pink tuff	1.4	7.7	1.48	19.1

per se is not of specific interest for analysing the mechanical effect of blasting; rather, the stress wave and piston effect of gases are of particular interest from the research point of view.

The mechanical effect of a blast depends mainly on the parameters selected for drilling-blasting operations, hook-up schemes and the properties of the massif. Unfortunately, these factors are not always given proper attention. Very often, the blast effect is evaluated individually under different conditions, giving difficult-to-compare relative quantities. The process of rock breakage has been examined by individual researchers in a blast-induced field of stresses in optically active materials. They attempt to judge the mechanism of breakage of rocks by blasting based on the rate of movement of crack equal to 0.4 C_p (C_p—velocity of propagation of longitudinal wave). In reality, such an interrelation cannot serve as a unique evaluation because of the dissipation of blast energy in time and space.

The various facets of the breakage process by blasting can be studied more comprehensively in laboratory models. In such an approach to the problem, the first question concerns the parameters of the stress wave and the second the dynamics of the processes of crack formation and crushing. To date the field of stresses induced by blasting a spherical charge has been studied more elaborately [7]. However, the modern technology of rock breakage by blasting envisages the use of shothole and blasthole charges. During a blast of a charge in a blasthole, the products of explosion move along the hole. The energy transmitted to the surrounding medium in a given section of the hole depends on the quantity of explosive per unit length, energy of the entire charge and progress of blast in time [28]. In this context, the need arose for conducting additional experiments so as to record the stresses that develop in the massif during blasting of a drillhole charge.

Three gauges were installed in the model block according to the layout shown in Fig. 26. The gauges were placed at a distance of 38 mm ($r/r_0 = 31.6$) from a charge of 2.4-mm diameter and length of 370 mm. The stress wave parameters are given in Table 19.

Table 19

Site for stress monitoring	r/r_0	Delay in arrival of wave pulse, µs	Duration of wave pulse, µs		σ_{max}, MPa
			Growth phase	Decay phase	
D-1	31.6	30	75	100	19 6
D-2	31 6	60	100	200	9.8
D-3	31.6	90	50	75	4.9

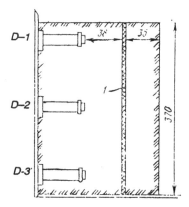

Fig. 26. Location of charge 1 and gauges D-1, D-2 and D-3 in the block

The wave pulse was shifted due to the detonation of the drillhole charge. The incoming wave was recorded by the upper (D-1 gauge), middle (D-2 gauge) and lower (D-3 gauge) points of the bench. The duration of stress wave pulse also differed; at point D-1, it was equal to 175 μs, at D-2 300 μs and D-3 125 μs.

Based on the experimental data and taking the velocity of wave of disturbances as approximately 1200 m/s and the rate of movement of crack as 210 m/s at a distance equal to 31.6 times the radius of charge (see Fig. 16), it was observed that the wavefront advanced 5 times or more than the head of the crack. In this case, the maximum stress in the wave exceeded the static threshold of material strength approximately eight times. Consequently, for breakage it is not enough for high stress levels to be induced; sufficient time is required for attaining the critical deformations.

It is obvious that time is an important factor that determines the breakage process. The duration of a blast effect during breakage of ore (rock) consists of the duration for which the wave pulse exists t_B and the duration of the piston effect of gases t_P, i.e.,

$$t = t_B + t_P. \tag{2.21}$$

The specific feature of the phase of piston effect of gases is the dynamic compression of the medium, as determined by time t_1, and growth of cracks prior to the disintegration of the body into fragments, indicated by time t_2:

$$t_P = t_1 + t_2. \tag{2.22}$$

An analysis of the experimental data revealed that while evaluating the rate of crack movement in the massif it is necessary to exclude time

t_1, lest relatively lower values of v_T be obtained. The rate of movement of a crack does not depend on the characteristic duration of dynamic compression t_1, which is established from the oscillogram recording the moment of breakage of the graphite rod situated in the first row from the charge.

A solution to the problem of partial participation of the blast wave and piston effect of gases in breakage is based not only on the results of experimental investigations, such as parameters of stress wave pulse, but also on the very breakage of the massif itself.

The time-dependent characteristics of wave pulse and durability of body subjected to a blast of single charges in model blocks are given in Table 20. In a zone limited by 5–60 times the charge radius, the value of the ratio of durability of body to the duration of stress wave pulse varies within 1.2–4.3. From this, it may be concluded that cracks in macroscopic volumes are formed during the phase of piston effect of gases, even without the existence of the stress wave. Evaluation showed that at the moment of formation of a macrocrack at a distance equal to 60 times the charge radius, the wave of disturbances would be at 6.6–7.9 m from the leading portion of the crack in the model. In actual conditions, this distance is equal to 40–53 m.

The computation of time-related contribution of the blast wave and piston effect of gases in rock breakage is of considerable interest. An analysis of experimental data (see Table 20) showed that with an increase

Table 20

No of block	Charge diameter, d_c, mm	Charge length, l_c, mm	r/r_0	Duration of wave pulse, t_B, µs	Durability of body, τ_d, µs	τ_d/t_B
4	2	160	5	60	90–100	1 5–1 67
9	2.2	120	12	80	130–140	1.62–1 75
87	1 75	180	26	120	160–180	1 33–1 50
94	1.75	180	35	160	250–280	1 56–1 75
24	2 2	190	36	160	250–300	1 56–1 87
31	2.2	170	40	280	340–350	1.21–1.25
141	2.0	180	40	170	340–350	2–2 06
175	1.75	180	42	150	370–400	2.46–2 66
67	1 75	180	42	180	370–400	2 05–2 22
174	2 0	180	42	170	370–400	2 18–2 35
81	1 75	180	43	130	380–420	2 92–3 23
85	1 75	180	47	180	480–500	2 67–2 78
92	1.75	180	47	180	480–500	2 67–2 78
95	1.75	180	47	170	480–490	2 82–2 88
59	2 0	180	50	150	500–600	3.33–4.0
174	2.0	180	60	200	750–860	3 75–4 3

ın distance from the blast site, the duration of wave pulse and piston effect of gases as well as the durability of the body increased. However, at a relative distance of 5–60, the time-related contribution of the wave in breakage was found to decrease from 67 to 23%, while the piston effect of gases increased from 33 to 77%. Thus, for example, when r/r_0 varied within 5–60, duration of existence of the blast wave pulse $t_B = 60$–200 μs, duration of the piston effect of gases $t_p = 30$–600 μs and durability of body $\tau_d = 90$–860 μs. In this case, the time-related contribution in breakage reduced from 67 to 23% and the piston effect of gases increased from 33 to 77%.

With increase in duration of the wave pulse as it moves away from the blast site, the amplitude of stresses becomes less (see Fig. 8). The fraction of energy remaining in the wave is about one-hundredth part of the energy of the entire charge. For example, at a relative distance of 40, it is equivalent to 0.0327–0.0530 or 3.3–5.3% (see Table 13).

The partial participation of the blast wave and piston effect of gases is assessed according to their contribution to the breakage process expressed by micro- and macroscopic volumes.

The work done to form a unit area of crack γ_0 in the wave phase of the blast effect can be determined by using Griffith's energy criterion. The quantity γ_0 and maximum stress σ_{max} in the compressive wave are related by

$$\gamma_0 = \frac{\pi(1 - \mu^2)\, \sigma_{max}^2 l_0}{2E}, \qquad (2.23)$$

where μ—Poisson's ratio;

E—elasticity modulus of the material; $E = (0.8$–1.0$)\, 10^3$ MPa (model);

l_0—average size of lump (of optimal fraction),
l_0—0.005 m.

Substituting the numerical values in expression (2.23), we find that with an increment in distance from blast site, the work done to form a unit area of crack decreases. For example, at a relative distance of 5, $\gamma_0 \approx 1.48$ J/m^2 ($\sigma_{max} = 140$ MPa, see Fig. 8) and at $r/r_0 = 40$, $\gamma_0 \simeq 0.03$ J/m^2 ($\sigma_{max} = 20$ MPa). This indicates that the zone of fractional participation of a wave in breakage is very limited, as the bulk of the blast energy is diffused near the charge during plastic deformation of the medium.

Processing the data on work done to form a unit area of crack, the deformations can be calculated by the formula

$$\varepsilon = 2\gamma_0/(\sigma_{max} l_0). \qquad (2.24)$$

The evaluation conducted so far shows that at relative distances of 5 and 40, the deformations are found to be equal to 4.2 and 0.6 mm respectively. However, as soon as the macrocrack forms, the initial stress field changes. At the initial stage of blast effect, radial cracks extending up to 2–3 times that of charge radius are formed in the contour zone. The results of earlier studies also confirm the localised failure in the contour zone of the drill hole [12].

The investigation of crack formation processes and crushing showed that the stress wave advances much farther than the failure front. During the period of existence of wave pulse macrocracks did not form even at a distance of 4 times the charge radius from the blast site. This implies that failure occurs at a relatively later stage of the process, when the medium has undergone considerable deformation.

The dynamics of the processes of crack formation and crushing is determined essentially by the extent of movement of ore (rock). Under the conditions of breakage by blasting in narrow stopes, for example in the mining of a vein type deposit, the surrounding massif impedes the movement of ore from the moment stress σ_T is attained and therefore the resistance to failure increases. On the contrary, major defects help in the formation of zones of leading (advancing) failure. The foregoing is recorded in experiments 45, 47, 50, 59 etc. (see Table 11).

An evaluation of results of complex investigations of stress-wave parameters, the process of crack formation and of crushing enabled a quantitative evaluation of fractional participation of the blast wave and piston effect of gases in the breakage process. Thus, during the phase of piston effect of gases radial cracks develop and also branch off. The fractional participation of the blast wave is mainly seen in the preparation of the massif to break (crushing), and quantitatively is about 10–20%. However, the mechanical effect is observed to be quite diverse during the phase of wave effect of the blast.

The contribution of the blast wave and piston effect of gases induced by blasting a unit charge can be schematically represented by means of the resultant fragmentation (quantitatively) in the following manner.

In breakage by blasting, zones of volumetric failure and crack formation are distinguished even though a clear boundary does not exist between them. In the first zone, the high-intensity wave mainly influences the fragmentation of rocks. In the zone of crack formation, fragmentation in microscopic volumes is related to the action of elastic wave (fractional participation 10–20%) while in macroscopic volume it is associated with the piston effect of gases (fractional participation 80–90%).

In the zone adjoining the charge, volumetric failure caused by the high-intensity wave occurs. The major part of the wave energy is expended on plastic deformation of the massif. It is better to improve the degree of utilisation of blast-wave energy to break the rock. These aspects are confirmed by results obtained by Baranov [5], who gives the key directions for optimising the processes of ore preparation according to the indices of their power intensity. The effect of blast loading on the physical and technological characteristics of ferruginous quartzite had been shown by Repin [8] and attests to the loss of strength in ore at the initial state itself.

The partial participation of the piston effect of blast products in fragmentation is considerably higher and quantitatively equals 80–90%. The energy of the piston effect of gases is mainly spent on moving ore (rock), on crack formation (fragmentation) and scatter of fragments. Thus the potential exists to improve the degree of utilisation of blast energy during the stage of piston effect of gases. For example, an evaluation of comparative results of breakage by blasting using igdanite and detonite draws attention to the relatively higher degree of fragmentation when igdanite is used. Detonite and igdanite, when blasted, release 835 and 750 l/kg of gaseous products respectively. If detonite is replaced by igdanite, the charge weight need be increased by only 10% in order to maintain the volume concentration of the gaseous products released by detonating the charge. However, the density of igdanite, when pneumatically charging drill holes or shot holes, is comparatively more and is equivalent to 1–1.25. Therefore, the specific consumption of gases in breakage increases, as a result of which better fragmentation of the massif is achieved. Concentrated gaseous products of explosion enhance the mechanical effect of a blast.

The established level of partial participation of the blast wave and piston effect of gases in rock breakage enables an evaluation of the explosives as well as of the work done by the blast and also optimisation of the parameters of breakage by blasting, with sufficient accuracy for practical use.

3

Model Investigations of Parameters of Rock Breakage by Blasting

3.1 Experimental Study of Breakage Process in Blasting and Recording of Stresses

The parameters of rock breakage by blasting as applied to the conditions of a base metal mine were studied by the method of equivalent materials (see Table 3, Composition III).

The accuracy of an experiment depends on the selection of scale and boundary conditions. The scale of modelling is dependent on two criteria—minimum strength of equivalent material and critical diameter of explosive charge. If the material is weak, the fragments which form after the blast will be intensively crushed against the walls of the model, distorting thereby the experimental results. The compressive strength of equivalent material—approximately 1.5 MPa—happens to be the boundary below which an error might occur. Therefore, it is advisable to select the scale of modelling as 1:50–1:20 in the ores and host rocks under study (see Table 1).

The critical diameter that ensures stable detonation of TEN in glass tubes is equal to 1.5 mm. A charge of TEN having a density of 1 g/cm^3 is used in the model; in actual commercial explosives (ammonites and grammonites), the charge diameters from the model to the actual and vice versa can be recalculated approximately using relationship (1.8). In such a case, it is important to ensure the equivalence of specific energy of explosive charges in the model and the actual.

During experimentation each block of the model of 500 × 500 × 400 mm was compacted on a vibrating table. Physicomechanical properties of the material used in the model were determined by testing six cubes with an edge dimension of 7 cm and six octahedral specimens. Elasticity modulus, velocity of longitudinal wave and Poisson's ratio were determined by testing a prism of 7 × 7 × 14 cm. After the model block attained the required strength, it was placed in the blast stand and, using a special device, firmly clamped in place. Metallic rods of 5-mm diameter, affixed to the model earlier, permit drill holes to be hung from them in

76

accordance with the parameters of field blasts and the selected scale of modelling.

A block diagram of the model is shown in Fig. 27. In the block (1), drill-hole charges of TEN (2) were blasted. Charges were initiated with a drop of lead azide (detonator) applied on the constantan resistance bridge. Detonators were fired from a 27-V circuit supplied by the exploder (3). In the models, extraction was by horizontal or vertical slices towards the compensation cavity. The blast-induced stresses were registered by means of the bridge (6), oscillograph (5) and camera (4). Working gauges (7) were oriented at 90° with reference to the charge axis before pouring material into the block and were affixed to the metallic frame of the model by two wooden screws.

The processing of experimental data included:

(1) Computation of specific consumption of explosive in primary: breakage* and volume of broken material beyond the perimeter of slice being blasted; (2) Establishing the granulometric composition of material and yield of oversize material. Additionally, the maximum stresses at the sides of the room, specific pulse, duration of blast wave effect etc., were determined from the oscillogram of the blast-wave pulse. The entire broken material was subjected to sieve analysis. Coarser fragments (of more than 40 mm size) were sorted manually.

A series of blasts was conducted in equivalent material models to study the features of fragmentation of ore as well as host rocks and the effect on dilution, taking into account the requirements of similarity of strength and elastic properties of material in model and rocks (ore) in the experimental district and satisfying the chosen scale of modelling.

The first series was conducted to obtain the relationship between specific consumption of explosive and yield of oversize material as well

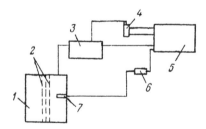

Fig. 27. Block diagram of the model. (Legend given in text.)

---

* The phrase 'in primary breakage' has been deleted hereafter because automatically assumed, unless specifically contrasted with explosive consumption in secondary breakage—Translator.

as dilution in the case of row-wise long-delay blasting with a value of LLR of 1.5 m. The results of these experiments are given in Table 21.

A second series of experiments was conducted to obtain the relationship between specific explosive consumption and yield of oversize material in a row-wise long-delay blasting and LLR of 2 m. The results of this series are shown in Table 22.

The results (Tables 21 and 22) indicate that row-wise long-delay blasting is distinguished by large yield of oversize fragments and high rate of dilution; therefore, another series of experiments was conducted using short-delay blasting.

The objective of the third series of experiments was to determine the relationship between specific explosive consumption and yield of oversize fragments as well as dilution at LLR of 3 m in row-wise short-delay blasting. Three slices of 18 cm (9 m in actual conditions) total thickness were broken by firing three rows of charges with delay intervals of 3.5 ms (25 ms in actual). The results of these experiments are given in Table 23.

A fourth series of experiments became necessary to obtain comparative results and were conducted by row-wise short-delay blasting at LLR of 2.4 m. Three slices of 18 cm (7.2 m in actual) total thickness were blasted by three rows of drillhole charges with 4 ms delay interval (25 ms in actual). The results of this series are presented in Table 24. It was observed that the yield of boulders decreased with a reduction in the drillhole grid (from 3 to 2.4 m in actual). In subsequent experiments, therefore, charges of small diameters were used so as to make the grid denser. Such a measure enabled a study of the granulometric composition of broken ore obtained by the two methods of blasting—row-wise with long delay and short delay.

The granulometric composition of broken ore obtained by blasting a row of drillhole charges at LLR of 1.5 m with long delays is given in Table 25.

The granulometric composition of broken ore obtained by blasting three rows of drillhole charges at LLR of 1.5 m with short delays is given in Table 26.

The condition of the model blocks was generally found to be satisfactory during investigations of the parameters of blast on three-dimensional models so chosen as to facilitate quantitative assessment of the results of each experiment. Failure of the solid massif varied depending on the selected breakage parameters and the method of blasting. This was recorded in each experiment. In the experiments conducted with short-delay blasting, sockets were found on the walls of the stoping room.

Adoption of the method of equivalent materials with recording of stresses enabled obtainment of experimental data for comparative evaluation of the parameters of breakage by blasting.

Table 21

Index	Serial number of blocks													
	39-a	38-a	36-a	37	36-b	38-b	39-b	36-c	38-c	39-c	36-d	39-d	36-e	39-e
LLR, mm	50	50	50	50	50	50	50	50	50	50	50	50	50	50
Slice length, mm	400	400	400	400	400	400	400	400	400	400	400	400	400	400
Slice width, mm	200	200	200	200	200	200	200	200	200	200	200	200	200	200
Drillhole length, mm	400	400	400	400	400	400	400	400	400	400	400	400	400	400
Spacing between drill holes, mm	50	50	50	50	50	50	50	50	50	50	50	50	50	50
No. of drill holes in a row	5	5	5	5	5	5	5	5	5	5	5	5	5	5
Drillhole diameter, mm	1 5	1 5	1 7	1 7	1 9	2 0	2.1	2 2	2 4	2 5	2 6	2 9	2.9	2 9
Explosive consumption for breakage, g	4	4	5	5 4	67	70	76	83	10.1	11	12 1	14.3	144	14 5
Specific consumption of explosives, g/dm³	1	1	1 25	1 37	1 7	1 76	1 9	2 08	2 53	2 76	3.03	3 58	3 6	3 61
Yield of oversize lumps, %	36	39	16	13	12	14	9	7	7	6	6	4	4	3
Dilution, %	6	0 6	11	12	14	10	5	13	25	19	16	28	18	31
Yield of ore per 1 dm of drill hole, dm³	0.2	0 2	0 2	0 2	0 2	0 2	0 2	0 2	0 2	0 2	0.2	0 2	0 2	0 2
Scale of modelling	1.30	1 30	1.30	1 30	1 30	1 30	1 30	1 30	1 30	1 30	1.30	1.30	1 30	1.30
No. of experiments	2	1	2	5	2	2	1	2	1	2	1	1	1	1

Table 22

Index	\	Serial number of blocks												
	44-a	43-a	41-a	42-a	42-b	43-b	42-b*	41-b	44-c	43-c	42-c	41-c	44-d	43-d
LLR, mm	50	50	50	50	50	50	50	50	50	50	50	50	50	50
Slice length, mm	400	400	400	400	400	400	400	400	400	400	400	400	400	400
Slice width, mm	200	200	200	200	200	200	200	200	200	200	200	200	200	200
Drillhole length, mm	400	400	400	400	400	400	400	400	400	400	400	400	400	400
Spacing between drill holes, mm	50	50	50	50	50	50	50	50	50	50	50	50	50	50
No. of drill holes in a row	5	5	5	5	5	5	5	5	5	5	5	5	5	5
Drillhole diameter, mm	1 5	1 5	1 7	1 7	1 9	2	2.1	2 2	2 4	2 5	2 6	2 9	2 9	2 9
Explosive consumption for breaking a slice, g	4	4	5	5 4	6.7	7	7 6	8 3	10 1	11	12 1	14 3	14 4	14 5
Specific consumption of explosives, g/dm³	1	1	1 25	1 37	1 70	1 76	1 9	2.08	2.53	2.76	3 03	3 58	3 60	3 61
Yield of oversize fragments, %	45	39	16	13	12	14	12	8	9	11	10	6	6	6
Yield of ore per 1 dm of drillhole, dm³	0 2	0 2	0 2	0 2	0 2	0 2	0.2	0 2	0 2	0 2	0 2	0 2	0 2	0 2
Scale of modelling	1:40	1 40	1 40	1 40	1.40	1 40	1:40	1 40	1.40	1.40	1 40	1 40	1 40	1 40
No of experiments	1	2	2	2	5	1	1	2	1	2	2	1	1	1

* *Sic*—General Editor

Table 23

Index	Serial number of blocks									
	17	20	18	13	15	16	8	9	4	19
LLR, mm	60	60	60	60	60	60	60	60	60	60
Slice length, mm	400	400	400	400	400	400	400	400	400	400
Slice width, mm	240	240	240	240	240	240	240	240	240	240
Drillhole length, mm	400	400	400	400	400	400	400	400	400	400
Spacing between drill holes, mm	60	60	60	60	60	60	60	60	60	60
No. of rows	3	3	3	3	3	3	3	3	3	3
No. of drill holes in a row	5	5	5	5	5	5	5	5	5	5
Drillhole diameter, mm	1.8	1.8	1.9	2.1	2.1	2.1	2.2	2.5	2.6	2.8
Explosive consumption for breakage, g	18	18.5	18.6	23.8	23.8	23.8	25.9	34.6	35.7	40.3
Specific consumption of explosives, g/dm³	1	1.06	1.09	1.4	1.4	1.4	1.5	2	2.1	2.37
Yield of oversize fragments, %	34	26.5	17.9	18	17	11.6	13	8.9	8.2	8.4
Dilution, %	3	5.7	3	6.4	4.8	5	5.2	10.3	15.2	16.3
Yield of ore per 1 dm of drill hole, dm³	0.3	0.3	0.3	0.3	0.3	0.3	0.3	0.3	0.3	0.3
Delay interval, ms	3.6	3.6	3.6	3.6	3.6	3.6	3.6	3.6	3.6	3.6
Scale of modelling	1:50	1:50	1:50	1:50	1:50	1:50	1:50	1:50	1:50	1:50
No. of experiments	2	2	1	1	1	1	2	1	2	2

Table 24

Index	Serial number of blocks									
	27	30	28	23	25	26	18	19	24	29
LLR, mm	60	60	60	60	60	60	60	60	60	60
Slice length, mm	400	400	400	400	400	400	400	400	400	400
Slice width, mm	240	240	240	240	240	240	240	240	240	240
Drillhole length, mm	400	400	400	400	400	400	400	400	400	400
Spacing between drill holes, mm	60	60	60	60	60	60	60	60	60	60
No. of rows	3	3	3	3	3	3	3	3	3	3
No. of drill holes in a row	5	5	5	5	5	5	5	5	5	5
Diameter of drill hole, mm	1.8	1.8	1.8	2.1	2.1	2.1	2.2	2.5	2.6	2.8
Consumption of explosive for breakage, g	18	18.5	18.6	23.8	23.8	23.8	25.9	34.6	35.7	40.3
Specific consumption of explosives, g/dm^3	1	1.06	1.09	1.4	1.4	1.4	1.5	2	2.1	2.37
Yield of oversize fragments, %	29	22.8	14.5	16.4	13.5	5.8	12.3	6.8	8	5
Yield of ore per 1 dm of drill hole, dm^3	0.3	0.3	0.3	0.3	0.3	0.3	0.3	0.3	0.3	0.3
Delay interval, ms	3.8	3.8	3.8	3.8	3.8	3.8	3.8	3.8	3.8	3.8
Scale of modelling	1:40	1:40	1:40	1:40	1:40	1:40	1:40	1:40	1:40	1:40
No. of experiments	2	2	1	1	1	1	2	2	1	2

82

Table 25

Specific explosive consumption for breakage, kg/m³	Yield (%) of fractions, mm				
	< 50	50–250	250–400	400–500	> 500
1 15	29 2	29 7	8 2	5 8	27 1
1 79	41 6	34 3	7.8	4 5	11 8
2 08	50	35 7	5	2 5	6 8
2 68	51	37	1 9	3 5	6 6
3 03	41 6	39 5	7 5	3 4	8
3 6	45 4	35	12	4	3 6

Table 26

Specific explosive consumption for breakage, kg/m³	Yield (%) of fractions, mm				
	< 50	50–250	250–400	400–500	> 500
1 07	16	39 9	11.7	27 9	4 5
1.4	18	48 2	17.8	16	—
1 47	20.5	45 7	16 2	16 4	1 2
1 7	25	51 8	12 8	10 4	—
1 8	25 5	56	10	8 5	—
2 09	26	54 3	13 7	6	—
2 37	27	52 3	16 3	4 4	—

3.2 Relationship between Breakage Parameters, Quality of Ore Fragmentation and Stability of Country Rocks

That the specific explosive consumption in breakage is one of the key parameters determining the quality of ore fragmentation is widely known. It appears that this may be solved simply by increasing or decreasing the specific consumption of explosive. However, with any such increase more country rocks get broken, thus diluting the ore. In recent years, requirements regarding quality of fragmentation have increased because of deployment of free-steered loading-hauling equipment in metal mines and subsequent autogenous grinding of ores in mills. Under such conditions it is important to increase the yield of medium-size fragments while breaking a minimum quantity of country rocks.

In the experiments conducted, the diameter of the drill holes was changed without varying their pattern to assess the influence of specific explosive consumption on the quality of fragmentation and dilution. The functional relationships between the chosen parameters and results of breakage by blasting were determined based on experimental data using mathematical statistics (see Tables 21 to 26).

Dilution R_* depends on the specific explosive consumption q, as

$$R_* = 12.5q - 1.07q^2 - 5.3. \tag{3.1}$$

Equation (3.1) is graphically shown in Fig. 28. The relationship between

83

the yield of oversize lumps K (standard fragment of 500 mm) and specific consumption of explosive in the case of row-wise long-delay blasting is also given in the same figure.

$$\text{at LLR of 1.5 m,} K = 32.8q^{-1.7}; \tag{3.2}$$
$$\text{at LLR of 2 m,} K = 33.2q^{-1.3}. \tag{3.3}$$

In row-wise blasting, the yield of oversize lumps increases accompanied by considerable breakage of the solid massif.

In the case of firing three rows of drillhole charges using short delays, the solid massif fails to a lesser extent (see Tables 23 and 24). The relationship between conditional dilution and specific explosive consumption in the case of short-delay blasting is of the form

$$R_* = 19.7q - 3.7q^2 - 13.9. \tag{3.4}$$

Equation (3.4) is graphically represented in Fig. 29. The curve showing the dependence of yield of oversize lumps on q in the case of short-delay blasting is also shown in the same figure. The yield of oversize lumps K is determined from the equations:

$$\text{at LLR of 3 m,} K = 32.3q^{-1.7}; \tag{3.5}$$
$$\text{at LLR of 2.4 m,} K = 28.8q^{-2.2}. \tag{3.6}$$

This experiment demonstrates that the dependence of dilution on specific explosive consumption differs in the two methods of blasting (see

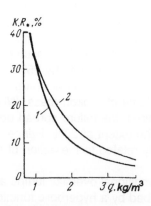

Fig. 28. Dependence of yield of oversize lumps K and dilution R_* on the specific explosive consumption q at LLR of (1) 1.5 m and (2) 2.0 m and long-delay blasting

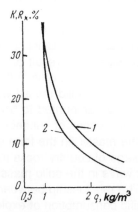

Fig. 29. Dependence of yield of oversize lumps K and dilution R_* on specific explosive consumption q at LLR of (1) 3 m and (2) 2.4 m, using short-delay blasting.

Figs. 28 and 29). Each specific method has its own parabolic relationship of conditional dilution and specific explosive consumption. The effectiveness of short-delay blasting is basically determined by the weight of the charge in a group, number of groups and delay interval. It is evident that the combination of these parameters in breakage by blasting significantly influences the quality of fragmentation and also the extent of breakage of enclosing rocks that cause dilution. Thus, in breaking three slices of 18-cm thickness (in real conditions 9 m) by firing three rows of drill holes using short delays (delay interval between rows 20–25 ms), the parabola grows even at a low rate of explosive consumption (see Fig. 29) and later, with an increase in explosive consumption, levels out. In the case of row blasting with long delays of a 5-cm thick slice (in real conditions 1.5 m), the parabolic curve extends (see Fig. 28). In short-delay blasting, a maximum dilution of 13.5% occurred at a specific explosive consumption of 3.3 kg/m^3. Further increase in explosive consumption with the given breakage parameters did not notably change the rate of dilution. A high increase in dilution rate caused by breakage of country rocks was noticed in row-wise ordinary blasting with augmented consumption of explosives.

Thus the quality of ore fragmentation depends mainly on the method of initiation of charges. In a blast of 43 g of TEN (5200 kg of ammonite 6ZhV in actual conditions) using short delays, the conditional dilution of ore amounted to 12% while in a row-wise blast of 9 g of TEN (430 kg ammonite 6ZhV in actual conditions) it was 18%. The condition of the room after blasting three rows of drillhole charges with 3.6 ms delay interval between rows (25 ms in actual conditions), was found to be satisfactory. In this series of experiments, sockets remained on the walls of the stoping room after blasting. In short-delay blasting, sockets were not seen but the solid massif failed. Blasting was done in a 20-cm wide room (6 m in actual conditions). As a result of row-wise blasting (at a specific consumption of explosive 1–3.6 g/dm^3) the room widened to 25 cm, i.e., by 25%. The failure of the solid massif within the zone of direct action of charges at the contact amounted to 5% of the total volume of broken mass. When drillhole charges of subsequent rows were fired, the failure spread deep into the massif (in the form of chipping) to the extent of 20%. Hence the breakage of country rocks in blasts is mainly related to the action of the blast wave in the solid massif.

The relationship between the yield of conditional oversize lumps and specific consumption of explosives is described by a hyperbolic function. With an increase in specific consumption of explosive and lower value of LLR, the higher the reduction in yield of oversize lumps. If at an LLR of 1.5 m, increase in the specific consumption of explosive from 1 to 4 kg/m^3 (by 4 times) causes a reduction in yield of oversize fragments from 32.8

to 2.5% (by 13 times), then at an LLR of 2 m and increase in the specific explosive consumption within the same limits causes a reduction in yield of oversize fragments from 33.2 to 5% (by 6.6 times).

The degree of reduction in yield of oversize fragments with an increase in explosive consumption differs according to the method of blasting. In short-delay blasting, an increase in specific consumption of explosive from 1 to 3.5 kg/m^3 reduced the yield of oversize fragments from 28.8 to 0.5% at LLR of 2.4 m and from 32.3 to 2.5% at LLR of 3.0 m. It is evident that in short-delay blasting a major part of the blast energy is expended on ore fragmentation and a small part in breaking the enclosing rocks.

A study of the principal laws of fragmentation confirmed the postulate that short-delay blasting is quite effective as it ensures the required quality of ore fragmentation with a minimum of dilution.

If the yield of oversize fragments is considered one of the major factors determining productivity in mining, then the quality of recovery depends mainly on the granulometric composition of ore, especially the yield of small fractions. Overcrushing of ore during extraction may increase losses. Such losses might possibly occur in blocks during the discharge of ores prone to consolidation. Moreover, autogrinding in mills has been widely adopted in recent years in ore beneficiation plants. According to the technological conditions of autogrinding, the composition of ore fines (fractions less than 50 mm) in the saleable ore should preferably be not more than 25%, thus making the quality requirements of fragmentation more stringent. In this context, it became necessary to study the conditional granulometric composition of the ore. For this a slice was broken by both methods—row-wise long-delay blasting and short-delay blasting of three rows of drillhole charges. The results of this experimental series are given in Tables 25 and 26. To analyse the yield of different fractions during breakage of the ore slice, the following groups of sizes were conditionally assumed: $K_1 < 50$; $K_2 = 50$–250; $K_3 = 250$–400; $K_4 = 400$–500 and $K_5 > 500$ mm. The relationship between quality of fragmentation (yield of fractions) and specific consumption of explosive in long-delay blasting was determined by the following relationships:

$$
\left.
\begin{aligned}
K_1 &= 108.4q - 22.1q^2 - 75.6; \\
K_2 &= 76.8q - 16.1q^2 - 45.2; \\
K_3 &= 3.6q^2 - 16.8q + 23.9; \\
K_4 &= 5.9q - 0.5; \\
K_5 &= 29.5q^{-1.5}.
\end{aligned}
\right\}
\tag{3.7}
$$

In breakage of a slice by short-delay blasting in three rows, the results obtained differed and the constraint equations were as follows:

$$\left.\begin{array}{l} K_1 = 11.2 + 7q; \\ K_2 = 51.2q - 11.6q^2 - 1; \\ K_3 = 1.3q + 11.9; \\ K_4 = 37.7q^{-2.5}. \end{array}\right\} \qquad (3.8)$$

Constraint equations (3.7) and (3.8) are graphically shown in Figs. 30 and 31 respectively. The experimental results (see Fig. 30) show that the yield of fine fractions was more in row-wise long-delay blasting. Thus, at a specific consumption of explosive of 1.15 kg/m³, the yield of ore fines (K_1) was 29.2%. With an increase in specific consumption of explosive, K_1 increased along a parabolic curve and the maximum yield (58%) of these fractions was obtained at a specific consumption of explosive of 2.4 kg/m³. The yield of fractions K_2 with a change in specific consumption of explosives from 1.15 to 3.6 kg/m³, varied along a parabolic curve from 29.7 to 35.0%, while the fractions K_3 and K_4 varied along a hyperbolic curve. However, the latter variation was distinguished by a relatively low yield. It was found that row-wise long-delay blasting is characterised by excessive uneven ore fragmentation. At 1.15 kg/m³ of specific consumption of explosives, 29.2% of fractions lower than 50 mm and 27.1% of fractions greater than 500 mm were obtained; the volume of medium fractions $(K_2 + K_3 + K_4)$ amounted to 43.7%.

The change in granulometric composition of ore obtained by short-delay blasting of three rows of drillhole charges is shown in Fig. 31. The delay interval between rows was taken as 3.6 ms (25 ms in actual). It

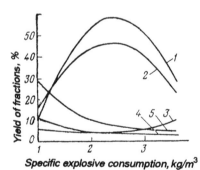

Fig. 30. Dependence of yield of fractions on specific explosive consumption in long-delay blasting and for class sizes of (1) < 50, (2) 50–250, (3) 250–400, (4) 400–500 mm and (5) > 500 mm

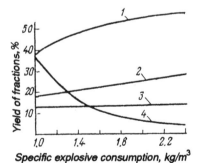

Fig. 31. Dependence of yield of fractions on specific explosive consumption in short-delay blasting and for class sizes of (1) < 50, (2) 50–250, (3) 250–400, (4) 400–500 mm.

was also found that this method of blasting results in relatively uniform ore fragmentation. In this case the yield of fractions K_1 was 18% at a specific explosive consumption of 1.15 kg/m³, while K_5 (larger than 500 mm) fractions were totally absent. With a change in specific explosive consumption from 1.15 to 2.4 kg/m³, the granulometric composition did not vary markedly. In this case the volume of average fractions $(K_2 + K_3 + K_4)$ was about 72–79.5%.

Using the correlation constraint equations (3.8), it was established that in the breakage of medium-hard ores by short-delay blasting of charges of 55–65 mm diameter in a drillhole grid of 1.5–1.6 m (at specific explosive consumption of 1–1.2 kg/m³), the yield of ore fines (fractions K_1) amounted to 16–20% while large ore lumps (fractions K_4) were about 25–37%. In these cases the dilution was minimal.

The aforementioned results are applicable to conditions of breaking ores with a strength coefficient of rocks ranging from 8 to 12 on the Protod'yakonov scale.The constraint equations on quality of fragmentation would differ in the case of harder ores. This divergence is all the more significant with an increase in hardness of ores and degree of discontinuities in enclosing rocks. A comparison of results of breakage by blasting obtained in models with those in actual practice enables such an assessment.

3.3 Relationship between Blast Wave Pulse and Stability of Enclosing Rocks and its Effect on Dilution

Much experimental data have been accumulated on the quality of extraction of minerals, especially of high-grade ores. Nevertheless, the phenomena of deformation and breakage of massif beyond the planned contour have not been adequately studied. Knowledge of the breakage mechanism of a solid massif enables selection of rational values of blast-breakage parameters that ensure stability of the enclosing rocks and the backfill.

The well-known theory of reflection of a compression wave from a free surface has been used to explain a specific situation of breakage. In so doing, the propagation of dynamic waves at the free surface of pillars between rooms, caused by the action of seismic waves, was well studied because pertinent to the chipping mechanism.

According to G.N. Kuznetsov, the process of chip formation consists of wave and quasi-static stages of failure of the exposed surface of the massif. At the wave stage the time interval distinguishing the moment of initiation of major (primary) crack from the initiation of detonation charge is equal to 4 and 40 μs respectively for models made of plexiglass and equivalent materials. A relatively high growth rate of primary cracks is characteristic of the wave stage. The initiation of primary cracks and their distance from the free face depend on parameters of stress wave and physicomechanical properties of rocks. The subsequent growth of primary cracks, culminating in chipping, occurs in the *quasi-static stage*. The author confirmed that the transverse wave propagating just behind the longitudinal one initiates the primary crack of a subsequent chip. To reduce the blast-induced seismic effects, Kuznetsov recommended that explosives be used which, when detonated, give a pulse of stress wave stretched in time; a changeover to short-delay blasting; and selection of a charge weight in a single delay such that the tensile stresses and deformations in the vicinity of the free face will not exceed the ultimate values.

Breakage of a solid massif is related to the chipping process. Kh.M. Aliev observed that the high-tensile stresses which cause chipping of material develop when a spherical wave propagating inside an elastic semi-space is reflected from the free boundary. With the phase of pressure drop diminishing at the incident wavefront, high-tensile stresses develop, forming the chip. This effect is used in ballistics of blasting, for example in solving the problem of dynamic strength. R. Schall showed that chipping occurred at that distance from the free face wherein the intensity of shock wave dropped by a quantity equivalent to the stress required for rupture. Thus for breakage, a stress-dependent time interval is required over which several chipping processes are initiated. The free face formed at the place of chipping alters the profile of stresses, as a result of which chipping does not occur completely at other points. Thus the viewpoint of R. Schall accords with the concept of G.P. Cherepanov, namely, that chipping occurs in a step-like curve of the 'ladder' or 'herringbone' type. According to K.B. Broberg, when the shock wave impinges on the reflected tensile waves at a specific distance from the free face, a tensile stress constant in time is induced. When this stress exceeds the static tensile strength, the material begins to fail. According to F.A. Baum, when a rarefied wave interacts with the shock wave (or wave of compression), tensile stresses develop at the free face, leading to the chipping of a portion of the rock. Nikiforovskii [16], while studying the process of chipping on the surface of glass films during a blast, established that rupturing cracks are initiated along surfaces perpendicular to the acting maximum tensile stress; the atomic layers become detached along these surfaces.

It is evident from the foregoing that the process of chipping occurring at the exposed faces in a blast breakage is due to the time-dependent wave criteria of breakage. The larger the amplitude of compressive stresses and the less the time earmarked for pressure drop behind the incident wavefornt, the higher the tensile stresses near the free face, as a result of which the breakage (chipping) process occurs more intensively.

However, the problem of chipping does not end with the description of any particular model of this process. The solution is related to the overall task of selection of rational parameters of blast breakage which will ensure the required quality of ore fragmentation while causing minimum breakage of the solid massif.

While investigating the parameters of blast breakage, it was established that the method of firing charges influenced not only the quality of fragmentation, but also breakage of the solid massif. This massif comprised mostly enclosing rocks. Breakage of enclosing rocks or backfill and undercutting of pillars between rooms is caused by the blast, which has been confirmed by actual practice in many metal mines.

The main task commonly encountered in a study of a solid massif is the state of the massif at the contact and investigation of its stability. The nature of development of stresses at the flanks of a stoping block (room) has evoked much interest in such investigations. Stresses were recorded using deep-seated transducers. In the experiments it was necessary to choose proper places for installation of the transducers. It is well known that in the existing methods of breakage by blasting, intensive breakage of rocks occurs up to a relative distance of 50-60 from the charge centre. Approximately at this distance considerable deformation is noticed in the deeper portions of the solid massif. In the usual method of blasting, breakage occurs at 1-5 cm in the model and 50–200 cm in the actual. When a transducer is installed at $r/r_0 = 20$, its working surface becomes exposed and the rod begins to vibrate at levels not characteristic of the process. At $r/r_0 = 30$, the transducer does not become exposed and stresses induced by the passing blast wave are recorded.

Experimental curves of time-dependent stresses (pulse) help in subsequent calculations and evaluation of blast-breakage parameters. Time-dependent stresses (Table 27) which developed at the flanks of the stoping room were recorded for both methods of blasting in models $50 \times 50 \times 40$ cm in size. It is evident from Table 27 that in row-wise ordinary blasting, alternating (in sign) stresses were induced at the sides of the room by the wave of compression and reflected tensile wave. This increased the amplitude of the blast wave. The maximum amplitude was recorded in row-wise ordinary blasting in blocks 38-c and 39-d. The magnitude of stress in the wave of compression amounted to 22 and 25.7 MPa, while that in the tensile wave was 3.7 and 3.8 MPa respectively. With an increase in pressure of the compression wave, a rise in pressure in the reflected tensile wave was also observed. Using the experimentally obtained stresses of compression and tension (see Table 27) recorded at the sides of the room, their interrelationship was established:

$$\sigma_{p_{max}} = -7.6 - 0.2\sigma_{c_{max}}. \tag{3.9}$$

Equation (3.9) is graphically depicted in Fig. 32. It can be seen from Table 27 that in experiments involving multirow (3 rows) short-delay blasting, tensile stresses are absent in blocks 28, 34, 35 and 101. Delay intervals between rows of drill holes of 1.5–2.9 mm in diameter were 4.5, 4.0 and 3.6 ms, depending on the scale of modelling (1 : 30, 1 : 40, 1 : 50). Considering the adopted scale of modelling, the delay interval in the actual is equal to 25 ms.

Thus the recorded graphs of stresses in the two methods of blasting differ notably. The difference lies mainly in the development of tensile stresses during long-delay blasting in the solid massif; these stresses were negligible or totally absent during short-delay blasting.

As mentioned above, with an increase in maximum compressive stress, the magnitude of tensile stress also increases. Therefore, with a reduction in r/r_0, in other words as the exposed surface comes closer, the effect of reflected tensile waves is more pronounced. Such a pattern is observed in row-wise long-delay blasting. Characteristic graphs of the stresses recorded at the sides of the room during blasts of drillhole charges with long delay (blocks 16, 36-e) and short delay (blocks 35, 101) are given in Fig. 33. In this series of experiments, stresses were recorded at a relative distance of 32–58 from the contour drill hole. In short-delay blasting, the blast pulse of the second or third row of drillhole charges was synchronised with the triggering channel of the oscillograph.

In the oscillograms pertaining to blocks 16 and 35, stress diagrams with two values of maxima were obtained. The first extremum was due to the effect of the longitudinal wave of compression, while the second

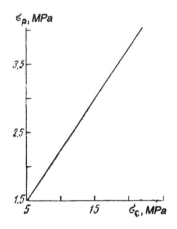

Fig. 32. Tensile stress versus compressive stress. Stresses were recorded in a long-delay blast at a relative distance of 35–52.

Table 27

Model block no	Stresses (MPa) developed at different time periods from the time of initiation of charges μs																	σmax, MPa
	35	70	105	140	175	210	245	280	315	350	385	420	445	490	525	560	595	
14	2	8.8	17.6	17	8	1.4	2	0.4	-1.6	-2.4	-3.2	—	—	—	—	—	—	19.1
16	4.1	17.2	15.8	4.1	1	2.3	0	0	0	-2.1	-2.1	-2.1	-6.6	0	0	—	—	17.6
28*	9.5	13	7	5.7	7	7.7	5.7	3.2	0.5	1.6	1.6	1.6	—	—	—	—	—	15.8
34*	13.3	14.6	12.3	6.6	4.6	2.7	1.3	3.3	3.3	4	3.2	3.6	2.3	0.3	0.2	0.2	1.5	20.6
35*	3.9	7.7	11.5	14.8	6.5	4.5	1.9	0.4	0	0	1.9	4.5	5.8	5.8	1.9	0.2	0.2	14.8
36-a	1.3	9.5	10.7	6.3	3.1	3.2	3.2	0.6	-3.2	-3.7	-3.7	-0.9	0	-0.9	0	0	0	11.4
36-b	1.7	3.9	5.6	5.9	5.6	4.9	3.9	3	1.7	0.9	0	0	-0.9	-1.4	-1.7	—	—	6.1
36-c	5.6	14	7.2	4	0.8	0.5	-0.2	-1.6	-2	-2	-1.6	-1.6	-2.4	-2.4	-1.6	-1.6	-0.8	14
36-d	2.5	3	5.1	5.1	6	5.1	3.4	0	0	-1.4	-1.7	-2.4	-1.7	-1.4	0.9	0	0	7.7
36-e	8	21.6	20	4.8	4.4	0.8	0	1.7	-0.3	-1.7	-2.4	-3.2	-3.2	-3.3	-3.6	—	—	22
37-a	3.3	10.6	6.2	4.9	0.8	-0.8	-2	-2.4	-3.2	-2.5	-2.4	-2.4	-2.4	-2.4	-2.4	-2	-1.6	11.5
37-e	7.8	11.5	4.9	2.6	1	1.2	—	-0.5	-1	-1.3	-0.9	-0.8	0	0	—	—	—	11.5
38-c	3.5	21.8	17.7	5.3	3.5	0	-3.7	-3.7	-2.1	-2.1	-1.2	-0.6	0	0	0	—	—	22
39-a	1.9	3.3	3.3	0.7	1	0	-1.4	-1.4	-1.8	-1.5	-1.3	-1.2	-0.7	-0.4	-1	-1.4	0	3.7
39-d	7.1	25.7	12.1	5.7	3.6	2.8	0	-2.9	-3.8	-2.8	-1.4	-0.7	0	0	0	—	—	25.7
39-e	6.2	14.8	17.2	13.7	10.9	7	3.9	1.8	0.2	0	-1.5	-2.3	-1.5	-1.2	-1.5	-1.6	-0.8	17.2
101*	1	3.4	2.5	3.7	3.7	1.2	0.3	0	0	0	0	0	0	0	0	0	0	6.1

*Short-delay blasting

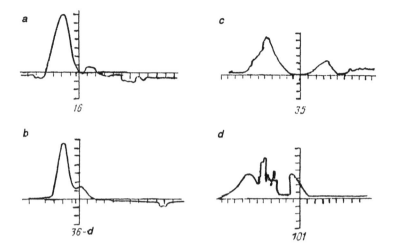

Fig. 33. Characteristic oscillograms of stresses developed in the solid massif during long-delay (a, b) and short-delay (c, d) blasting.

was due to the blast effect of the adjacent charge in the row. The duration of action of the blast wave pulse on the solid massif increased when an adjacent charge was blasted. In the oscillograms of blocks 16 and 36-e, at periods of blast wave pulse 233 and 254 μs respectively, the stresses due to the effect of the blast wave of the adjacent charge were observed after 134 and 195 μs, while the effect lasted for 99 and 59 μs, and the pressure amounted to 3 and 5.9 MPa. The blast wave pulse in the short-delay blasting method lasted for a relatively longer period. In the oscillograms of blocks 35 and 101, the stresses induced by blasting three rows of five drillhole charges in each were recorded. The delay interval between the rows was taken as 4.5 μs. The period of blast wave pulse in the oscillograms of blocks 35 and 101 was 580 and 300 μs respectively. The duration of the effect of the blast wave and stress depends basically on the relative distance between the place of recording and the charge. With an increase in r/r_0, the pulse period increases while stresses decrease. In the oscillogram of block 101, four extrema occurred when a group of drillhole charges placed in a single row were fired for breaking three slices. The fifth extremum was absent due to the misfiring of one of the charges (fourth from the transducer). It is obvious that in short-delay blasting even though the solid massif is situated in a relatively extended stress state, the stability is not disturbed due to the absence of tensile stresses. The tensile stresses at $r/r_0 = 32$–58 equalled 1.3–3.8 MPa. The dilution was found to be relatively higher (25–31%) in blocks 38-c, 39-d and 39-e, in which maximal tensile stresses were recorded (2.3–3.8 MPa).

Thus in row-wise long-delay blasting, the stability of a solid massif proves inadequate; it may break even at a low specific explosive consumption due to the development of tensile stresses at the contact. Based on the experimental data above and considering relationship (1.15), it was established that the tensile stress at the surface of the solid massif (under actual conditions) amounted to 40–160 MPa, which is significantly greater than the static strength of rocks.

In experiments conducted to study the parameters of blast breakage, the charge diameters were taken as 1.5–2.9 mm in a constant hole grid. It is obvious that along with variation in the charge diameter, r/r_0 also varies; hence it is difficult to establish a specific relationship between stresses developing in the enclosing rocks and the dilution. In such conditions the blast pulse or the specific pulse is a more convenient parameter for assessing the stability of enclosing rocks in row-wise long-delay blasting. The parameters of the stress wave established while modelling the breakage of ore by drillhole charges are given in Table 28. According to the experimental data σ_{max} and measured values of γ and C_p by solving equation (1.9) relative to v_{max}, we get the value of maximum velocity of particles. With an increase in charge diameter, the velocity of particles is given by

$$v_{max} = 101 d_c - 128. \tag{3.10}$$

Equation (3.10) is graphically depicted in Fig. 34.

The relationship between specific pulse and specific explosive consumption is given by the equation

$$I_0 = 43 + 49.7q. \tag{3.11}$$

This relationship is graphically shown in Fig. 35.

The dilution rates for breakage with long-delay blasting of drillhole charges are given in Table 29. The data of experimental investigations show that a linear relationship exists between the parameters under study. While computing the parameters of equations pertaining to the linear relationship by the method of least squares, the equation of a straight line is obtained that establishes the dependence of dilution rate on the scaled specific blast pulse,

$$R_* = 0.12 I_{sc}, \tag{3.12}$$

which helps in accurately describing the phenomenon under study. The relationship (3.12) is graphically shown in Fig. 36.

Hence the method of blasting markedly influences the breakage of enclosing rocks and ore dilution. Large-scale breakage of a solid massif in row-wise long-delay blasting is associated with inducement of alternating stresses. In the short-delay blasting method with rational breakage parameters, the tensile stress at the contact does not reach a value that

Table 28

Block no.	Drillhole diameter, mm	Specific explosive consumption, g/dm³	Delay interval between rows of drill holes, ms	Maximum pressure in wave of compression, MPa	Duration of wave of compression, μs	Specific pulse, N·f/m²	Maximum particle velocity, cm/s
13	2.1	1 4	—	14.3	301	133 6	86 3
14	2 1	1.7	—	19.1	279	171 3	115 3
16	2 3	1 78	—	17.6	351	168 4	106 3
28	1.9	1.09	3 5	15.8	335	138 4	95 4
31	2.2	—	—	16.2	284	190 7	97 8
34	2 3	2 1	4	20.6	241	163 8	124.4
35	1.9	1 51	4 5	14 8	297	167 4	89 4
36a	1 7	1.25.	—	11 4	305	121.3	68 8
36-b	1.9	1.7	—	6 1	345	115.1	36.8
36-c	—	2.08	—	14	242	112 2	84 5
36-d	2.6	3.03	—	7 7	323	117 8	46.5
36-e	2 9	3.6	—	22	233	171.9	132 8
37-a	1.8	1 37	—	11 5	183	101.9	69 4
37-b	1.8	1 37	—	12.1	222	97.7	73
37-e	1.8	1.37	—	11 5	220	108 1	69.4
38-c	2.4	2.53	—	22	205	172 9	132 8
39-a	1.5	1	—	3.7	320	34	22.3
39-d	2.9	3.58	—	25.7	208	216 9	155 2
39-e	2.9	3.61	—	17.2	300	255 8	103 8
101	1.8	1.35	4 5	7	—	61	42 3

Table 29

Model block no.	Specific pulse, N·f/m²	Scaled specific pulse (N·f/m²) 10³	Dilution, %
13	133.6	127 2	6
14	171 3	163.1	17
16	168.4	146 4	19
36-a	121.3	142.7	11
36-b	115.1	121 1	14
36-c	112.2	97.5	13
36-d	117.8	90 6	16
36-e	171 9	118.5	18
37-a	101.9	113 2	12
37-b	97 7	108.5	12
37-e	108.1	120 1	12
38-c	172.9	144	25
39-a	34	45 3	6
39-d	216 9	149.6	28
39-e	255.8	176 4	31

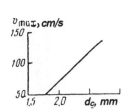 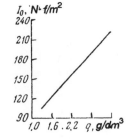 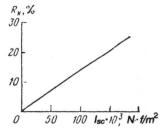

Fig. 34. Dependence of particle velocity on charge diameter.

Fig. 35. Dependence of specific pulse on specific explosive consumption.

Fig. 36. Dependence of dilution R_* on the scaled specific pulse I_{sc}.

could actually cause breakage, even when the duration of pulse is increased.

The effect of short-delay blasting of 3–4 rows of drillhole charges of 65 mm in diameter can be used in breaking ores. However, even in this case the delay interval between rows should be 20–30 ms. The range of delay intervals in the case of breakage in models is limited to 2.9–4.3 ms (Fig. 37). Beyond this range of delays, increased breakage of the solid massif is noticed. This should be kept in mind while designing and organising blast breakage in underground mines. Experimental data on the growth of stresses under different blasting methods are given in Table 30.

96

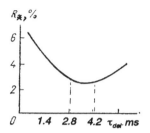

Fig. 37. Dependence of dilution R_* on delay interval τ_{del} between rows of drillhole charges.

Table 30

Time elapsed from beginning of firing charges, µs	σ_{max}, MPa, in	
	Row-wise long-delay blasting	Triple row blasting with delay interval between rows 3 6–3 8 ms
35	4.2	8.9
70	12.7	11.8
105	11	10.3
140	6.2	9
175	3.8	6 3
210	2	5
245	0 8	3
280	−0.3	2 3
315	−1.3	1 3
350	−1 7	1 9
385	−1.8	2.2
420	−1.4	3.2
455	−1.1	2.7
490	−0 9	2
525	−1	0 7
560	−0 5	0.1
595	−0 3	0 6

The ore is blasted in slices using drillhole charges. Multirow short-delay blasting happens to be a more productive method but sometimes it is advisable to adopt row-wise long-delay blasting. The seismic effect induced by long-delay and short-delay blasting differs notably. Even the mechanism of breakage of the solid massif differs.

The diagram of stresses developing at the contact in short-delay blasting is quite interesting (Fig. 38). During the firing of second-row charges the wave falls on the exposed face and interacts with the residual field of stresses, due to which compressive stresses may develop or may be totally absent. The resultant value of stresses depends on the absolute value of incident and reflected waves. In the breakage of ore by vertical or horizontal slices by means of long-delay blasting of drillhole charges, the enclosing rocks fail according to the scheme shown in Fig. 39. Failure in microscopic volume occurred in zone 1 adjacent to charge 2 due to the blast-wave effect. Under the effect of multiple repeated loading of the blast wave (due to breakage of successive slices) failure occurred in the macroscopic volume near the free face (3). At this juncture a wave pulse with two peak values was recorded—the first quantifying compressive stress and the second the tensile stress. Hence the appearance of tensile stresses near the free face and failure of the solid massif in the room (4) conformed to each other. This model of breakage of enclosing rocks (5) corresponds to row-wise long-delay blasting. Vibrations of a large amplitude in the enclosing rock massif and failure of the solid massif are characteristic of this method of rock breakage. Here the minimum tensile stress is proportional to the growth in compressive stresses. The selection of rational delay intervals is an important measure for controlling ore (6) fragmentation, while considering the stability of the solid massif.

As a result of these experimental investigations into the pattern of ore fragmentation and failure of a solid massif, a range of rational val-

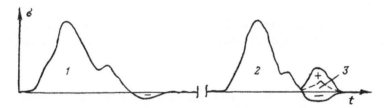

Fig. 38. Diagram of stresses developed in the solid massif during short-delay blasting.
1—wave pulse of first-row charges; 2—wave pulse of second-row charges; 3—stresses arising due to interaction of pulses of first and second rows.

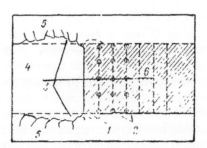

Fig. 39. Diagram indicating failure of a solid massif in row-wise long-delay blasting of ore (Legend given in text)

ues of blast breakage parameter could be established: charge diameter 55–65 mm; drillhole grid 1.5 × 1.5–1.6 × 1.6 m, delay interval between rows 25 ms with a deviation of ±20%. With these parameters it is possible to ensure the required quality of fragmentation, taking into account the requirements for autogenous grinding in crushers while mining hard and medium-hard ores (yield of fractions < 50 mm not to exceed 25%).

The relationship between the yield of oversize material and specific explosive consumption typical to each method of blasting was determined. In short-delay blasting, ore fragmentation is uniform compared to the ordinary method: the yield of average fractions increases by two times due to ore fines and oversize fragments.

A specific relationship between the dilution and specific explosive consumption characterises each blasting method. In long-delay blasting, failure of the solid massif begins at a specific explosive consumption of 0.5 kg/m^3, which in short-delay blasting is 1 kg/m^3.

The state and conditions of stability of a solid massif were studied. Large tensile stresses developed in the zone abutting the exposed face, when the ordinary blasting method was adopted. At a relative distance of 32–58, the stresses in the model were equal to 1.3–3.8 Mpa, and under actual conditions 40–160 MPa. In short-delay blasting, with a delay interval of 25 ms between rows, the tensile stresses were insignificant.

With an increase in specific blast pulse in row-wise long-delay blasting, the broken (chipped) volume of solid massif increased proportionately.

3.4 Experimental Study of Ore Breakage by Blasting Drillhole Charges with Presplitting of Narrow Stope Area

The data of the foregoing experimental investigations are pertinent to the breakage of ore in wide deposits. In underground metal mines, steeply dipping veins of average thickness are usually worked by shothole charges and very rarely by drillhole charges. The methods of breakage known to date are not very effective: firstly, the firing of shothole charges results in overbreakage of ore, a major part of which remains in the footwall due to the uneven surface; secondly, in drillhole charges the ore breakage is not uniform and the solid massif is broken.

The method of mining an ore vein by cutting slices along the strike using drillhole charges is rather well known. It involves drivage of a drill raise, drilling of parallel holes over the entire length of the slice, charging the holes and sequential blasting of charges from the cut holes to the contour holes [17]. Irregular fragmentation of ore and substantial breakage of enclosing rocks due to loss in blast energy directed towards irreversible deformation of the massif are the demerits of this method.

Another method of extracting minerals with long-delay preblasting of charges in perimeter drillholes is also known [1]. In this method, a screening zone is created by breaking pillars between drill holes of the perimeter row and subsequent firing of charges in the massif is done from the perimeter to the centre with delays between rows. The demerit of this method is the irregular ore fragmentation due to irrational utilisation of the blast energy of the perimeter charges.

The objectives of the present investigation were to reduce the yield of ore fines, to obtain even surfaces on the walls of the room, to increase the productivity of miners and to reduce the consumption of explosive materials for breaking ores in steeply dipping veins of average thickness.

In this context, the blast breakage was modelled in ores from a base object with a strength coefficient of 10–14 on the Protod'yakonov scale (Table 31).

To obtain comparable results, breakage was done with subsequent contouring and precontouring in blocks of models of $500 \times 500 \times 400$ mm. The scale of modelling was 1 : 25.

The results of breakage of a 2-m thick vein with subsequent contouring of the stope area in the model[1] are given in Table 32.

The results of breakage of a 2-m thick vein with precontouring of the stope area in the model are given in Table 33.

The breakage scheme of the vein (1) in a narrow stope is shown in Fig. 40. The perimeter holes (2) are situated at the contact at a distance L from each other and in the same slice a cut hole (3) is also drilled. The number of cut holes depends on the vein thickness and varies from one to three for a thickness between 1.4 and 3.6 m. Perimeter charges are fired simultaneously, as a result of which a cavity (4) is formed. Moreover, a disturbed zone (5) with cracks is formed, which determines the

Table 31

Index	Biotitic sandstones	Andesitic and dioritic porphyry	Composition III (see Table 3)
	Actual		Model
Compressive strength, MPa	100–140	120–170	4–6
Tensile strength, MPa	9–3	10–16	0.3–0.5
Density, g/cm³	2 5–2 6	2.5–2.8	2 5–2 6
Poisson's ratio	0 19–0 24	0 3	0 19–0.22
Elasticity modulus, 10^{-5} MPa	0 2–0.6	0 2–0 7	0 008–0.01
Velocity of longitudinal wave, m/s	3500–4200	2800–4600	650–750
Acoustic rigidity, 10^{-5} (N·f)	8 75–10.9	7 0–12 9	1 63–1 95

[1]S S. Gasparyan participated in this and subsequent experiments.

Table 32

Charge diameter d_c, mm	Delay interval τ_{del}, ms	Charge length l, mm	Specific explosive consumption, q, g/dm³	LLR, mm	Yield of oversize particles, K_{op}, %	Yield of ore fines ($<$ 50 mm) K_{of}, %	Depth at which enclosing rocks (ore) broke, H_p, mm	Dilution R_*, %	No. of experiments
1.7	0	400	1.29	60	11'	18.4	14	17.5	2
	1	400	1.29	60	8.6	16.8	13	16.2	3
	3	400	1.29	60	8.8	16.4	12.4	15.5	3
	5	400	1.29	60	8.7	16.4	12.5	15.6	3
2	0	400	1.95	60	9	20.6	16.5	20.6	2
	1	400	1.95	60	6.7	18.5	15.2	19	3
	3	400	1.95	60	7	18.4	14.5	18.1	3
	5	400	1.95	60	7.4	17.7	14.5	18.1	3
2.3	0	400	2.58	60	7.5	24	19.8	24.8	2
	1	400	2.58	60	6.6	20.1	17.9	22.4	3
	3	400	2.58	60	6.2	19.8	17.7	22.1	3
	5	400	2.58	60	6.4	19.8	17.3	21.6	3

Table 33

d_c, mm	τ_{del}, ms	l, mm	LLR, mm	q, g/dm³	K_{op}, %	K_{of}, %	H_p, mm	R_*, %	No. of experiments
1.7	0	400	60	1.29	11	18.4	14	17.5	2
	1	400	60	1.29	7.2	14.7	9	11.2	4
	2	400	60	1.29	6.8	14	7.7	9.6	4
	3	400	60	1.29	4.1	12.1	4.6	5.7	4
	4	400	60	1.29	5.8	13.2	6	7.5	3
	0	400	60	1.95	9	22.6	16.5	20.6	2
2	1	400	60	1.95	5.5	16.2	10.8	13.5	3
	2	400	60	1.95	4.8	15.6	9.6	12	4
	3	400	60	1.95	3	13.6	5.8	7.2	4
	4	400	60	1.95	4.4	14.7	7.5	9.3	3
	0	400	60	2.15	8.6	20.5	17.7	22.7	2
2.1	1	400	60	2.15	5.2	16.6	11.7	14.6	3
	2	400	60	2.15	4.3	15.9	10	12.5	4
	3	400	60	2.15	2.4	14.5	6	7.5	5
	4	400	60	2.15	3.8	15.8	7.8	9.7	4
	0	400	60	2.58	7.5	24	19.8	24.8	3
2.3	1	400	60	2.58	3.4	18.9	13	16.2	3
	2	400	60	2.58	2.4	18.2	12.2	15.2	4
	3	400	60	2.58	0.6	16.3	7.5	9.3	4
	4	400	60	2.58	1.7	18.2	9.8	12.2	3

room boundaries. Fragmentation is irregular in the case of ore breakage with post-splitting. In addition to the walls of the room having an uneven surface, ore fines are lost in the foot wall. In the case of breakage with presplitting, contrarily, fragmentation is uniform, the walls of the room remain even and, therefore, ore fines are not lost while drawing.

The proposed objectives are achievable through a novel method in which the cuthole charges are fired with a delay interval relative to the perimeter charges. This method distinctly differs from the already known method of firing perimeter charges and subsequent firing of cuthole charges [1]; the delay interval is found from calculations and also experimentally confirmed. The chosen delay interval ensures a corresponding enlargement of the surfaces of cavity B (see Fig. 40) and a sufficient number of growing microcracks in the ore being broken. The experimental data show that the cumulative effect of microcracks reduces the strength of the massif to 20%.

A screening cavity forms as a result of the movement of the massif under breakage towards the compensating area due to the piston effect of the products of explosion. The dimensions of the cavity depend basically on the chosen LLR. Experiments have shown that the ultimate width reaches 4.8 mm in the model under the parameters selected (in a blast of 2-mm diameter charges) and 120 mm in an actual mine (52-mm diameter charges).

The influence of the dimensions of the cavity needs to be emphasised. In the enlargement process of cavity surfaces up to $B = 0.05$–0.08 m, the slice being mined moved in the direction of the compensating area along LLR. Firing of the cuthole charge with a fixed delay interval was accompanied by a rise in kinetic energy of interacting particles of the ore slice being extracted. This is associated with accelerated movement of the ore in both the direction of principal compensating area and the cavity. Widening of the boundaries of slice movement and acceleration

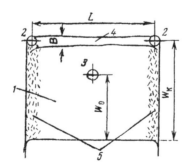

Fig. 40. Charge layout and cavity formation (Legend given in text)

of particles is accompanied by significant branching of cracks, leading to uniform ore fragmentation.

The experimental data show that breakage with presplitting, compared to post-splitting, enables a reduction in breakage of the solid massif of up to 40–50%, yield of ore fines 18–25% and oversize fragments 40–50%.

With the transition from small shot holes to drill holes and variation in the conditions of working of explosive charges in narrow stopes, it is necessary to consider the following important factors that influence the value of LLR. The collar of the shot holes drilled in the adjacent slice is situated near the top of the ore body, already disturbed by the preceding blast, while the bottom of the massif remains practically untouched; this is due to the zone of crack formation around the charge. The drill-hole charges, in contrast to the shothole charges, work in the ore massif undisturbed by blasting. In this case the zone of crack formation occurs at the ceiling of the stoping area, parallel to the drillhole charges. Due to the weakening of the massif by cracks, the strength of this part (ceiling) of the massif is relatively lower; it becomes possible therefore to drill holes taking into account the zone of crack formation, i.e., to increase the LLR of perimeter charges roughly by 1.5 times.

Given the foregoing, the effect of the LLR of perimeter holes W_K on the quality of fragmentation in the case of constant diameter of charges of 2 mm in the model (actual 51 mm) was studied. The LLR in the model was 45–80 mm (actual 1.13–2 m). The results of model investigations into the effect of W_K on ore fragmentation while working a 2-m thick vein with presplitting of the stope area are given in Table 34.

It follows from Table 34 that the condition for quality of ore fragmentation is

$$1.4W_o \leqslant W_K \leqslant 1.5W_o. \tag{3.13}$$

where W_o is the LLR of the cut hole.

In the case of deviation from boundary conditions (3.13), the quality of ore fragmentation might be worsened. If $W_K \ll W_o$, an ejection cone

Table 34

d_c, mm	τ_{del}, ms	l, mm	W_K, mm	q, g/dm^3	K_{op}, %	K_{ot}, %	No of experiments
2	3	400	45	2 6	0 3	48	2
2	3	400	50	2 34	0 9	29 3	3
2	3	400	55	2 12	2 1	20 5	3
2	3	400	60	1 95	3	13.6	4
2	3	400	65	1 80	9.5	12	3
2	3	400	70	1 67	21 5	10.2	3
2	3	400	80	1 46	46	8 1	2

may form in the cuthole charge and a 'blowout' occur in the perimeter charges. While if $W_K \gg W_0$, some amount of ore may be ejected by the cuthole charge.

The method using drill holes for working steeply dipping, medium-·thick veins of ore was developed consequent to laboratory modelling and development of rational values of parameters for drilling/blasting operations to ensure reduced yield of ore fines and oversize fragments, taking into account the stability requirements of the solid massif.

4

Field Verification of Results Obtained Through Modelling and Determination of Rational Values of Ore Breakage Parameters

4.1 Experimental Study of Ore Breakage Under Actual Conditions

An experimental section in a base metal mine was represented by hydrothermally altered talc-carbonaceous rocks. The strength coefficient of the rocks varied from 3 to 20 on the Protod'yakonov scale, with the predominant values being 8–12. The hanging wall of the ore body, within the limits of the experimental section, consisted of competent gabbro, while the footwall comprised peridotites. The metal values were very irregularly distributed in the ore and the limits of (economically feasible) mineable ore were determined based on sampling data. The elements of deposition of the ore bodies were not stable with the angle of dip mostly in the range 75–85°.

The varying stability of ore and waste caused by tectonic disturbances and development of cracks limits the field of application of conventional methods of working. In such conditions, abandoning part of the metal values at the contact or breaking the solid massif in the process of extracting ore may lead to a sharp fall in the average metal content. Efforts aimed at enhancing the quality and complete recovery of ores from the earth's crust have led to wider application of the method of working with cemented backfill. Additionally, the requirements imposed on blasting have also become more stringent. The specific characteristic of ore breakage should be studied taking into account the aforementioned peculiarities of working the deposit.

The design of experiments involving ore breakage under field conditions is associated with certain known difficulties. Such notwithstanding, 105, 85, 65 and 41-mm holes were drilled in the blocks of the experimental section. The various operations undertaken are shown in Table 35.

The ore was broken in one of the chambers in block 1/98 using drill holes of 105-mm diameter (Fig. 41), and in the other by shot holes of

Table 35

No of blasts	Charge diameter, mm	Volume of broken ore, T	Volume of ore drawn under observation, T	No of observations in a shift while drawing ore
15	105	32,930	19,870	249
162	41	5,470	3,120	41
7	85	4,750	4,627	26
28	65	14,273	14,000	129

41-mm diameter. During exploitation of this block, the tendency of the ore towards consolidation and uneven fragmentation was established. The large yield of oversize fragments sharply reduced labour productivity in ore withdrawal, while its overcrushing was conducive to consolidation. These factors led to the arching of broken ore lumps and hangups (arches) in the stoped out area. According to observations conducted over 249 shifts during ore withdrawal in the 1/98 block, the yield of oversize lumps was 14.3–16% at a specific consumption of explosive (ammonite 6ZhV) for direct breakage of 1.506–1.518 kg/m^3. The specific consumption of explosive for secondary breakage amounted to 0.204–0.218 kg/m^3.

It can be seen from Table 35 that during the process of blasting by drill holes of 105 mm, only 62% of the entire broken ore was drawn due to its consolidation. This led to discarding of the shrinkage method of working and also the method of working with breakage of horizontal slices and the switchover to breakage by vertical slices. In the method of breakage by

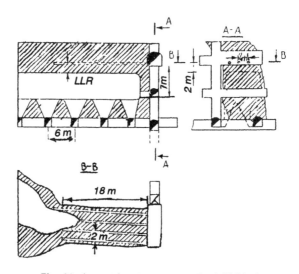

Fig. 41. Layout for mining ore in the 1/98 block

horizontal slices, the ore stored in the stope became consolidated under the repeated effect of blasts. Adoption of the method of breakage by vertical slices on the compensating cavity contributed to increased productivity and quality and rate of recovery. Because of the altered drillhole pattern it became necessary to constructively redesign the method of working. A local design for working block 2/5 was developed that considered using sublevel drifts (Fig. 42).

In block 2/5, the ore was broken using drill holes of 85 and 65 mm. The parameters of the block were: length 60 m, height 40 m and width equal to the width of the ore body. The pattern usually adopted was of parallel drill holes and very rarely of the fan type. The fan pattern was used only in the case of complex morphology of the ore body. In the process of working the first chamber of block 2/5, two series of experiments involving ore breakage by vertical slices were planned. Holes of 5–15-m depth were drilled by bits AK-85 and AK-65 along the rise side of the ore body using PT-36 drills [telescopic drills] from the sublevel drifts. In toto, 35 blasts were conducted, resulting in the breakage of 19,023 T and 18,627 T of ore was drawn. In all the experiments, the breakage parameters were recorded (blast scheme, charge weights etc.) and the yield of oversize lumps or coarse fragments (400–500 mm) was also determined.

The *first series* of experiments stipulated the determination of quality of ore fragmentation in row-wise long-delay blasting of charges in drill holes of 85 mm diameter. It may be noted that the drillhole patterns in

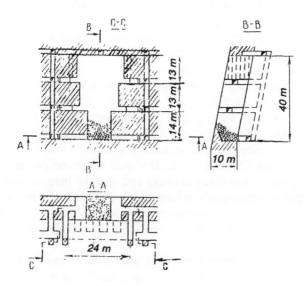

Fig. 42. Method of working with sublevel drifts

breakage by vertical slices is relatively flexible since the entire set of drill holes can be fitted in a better manner into the mineralisation contours. The technoeconomic indices of breakage by drillhole charges of 85-mm diameter were better compared to 105-mm diameter holes. However, the results obtained do not lie within the field of rational values. With a drillhole pattern of 1.8 × 1.8 m and at a specific explosive consumption for direct breakage of 1.12–1.17 kg/m^3 and for secondary blasting 0.19–0.21 kg/m^3, the yield of oversize lumps was about 15–16.2%.

While studying the patterns of ore fragmentation in the blocks of the model it was established that qualitatively better fragmentation was obtained by reducing the diameter of the holes; by compressing the grid the yield of oversize lumps reduced as did ore fines while the country rocks diluting the ore were minimal. The field of rational values of blast breakage parameters was determined: charge diameter 55–65 mm, hole pattern 1.5 × 1.5–1.6 × 1.6 m and rational delay interval between rows of drill holes 25 ms.

The *second series* of experiments envisaged investigation into ore breakage under field conditions using these parameters in order to compare results. The experiments involved the breakage of vertical slices on the compensating cavity by firing three rows of drillhole charges with 25 ms delay between rows. Holes were charged with ammonite 6ZhV cartridges of 60-mm diameter. The specific explosive consumption in primary breakage was regulated by altering the 65-mm diameter drillhole grid. Thus, besides changing the drillhole grid (LLR and drillhole spacing) from 1.2 to 2 m, the specific explosive consumption in primary breakage was varied from 0.7 up to 1.94 kg/m^3. Experimental results of the second series are given in Table 36. Taking into account the relatively better breakage results obtained by using drill holes of 65 mm, this method was introduced on an industrial scale.

The method of working with sublevel drifts was tested in field trials by specialists from Geological and Mine Survey Department of the base metal mine and also by a group of specialists from Armniprotsvetmet. Qualified workers were deployed in the ore breakage process. Sampling results of run-of-mine ore and saleable ore showed that the average rate of dilution in block 2/5 was 3.6%. This was achieved by using holes of smaller diameter, short-delay blasting and rational hole pattern. Increase in the yield of average-size ore lumps with simultaneous reduction in the yield of oversize lumps and ore fines was a key achievement in the mining of block 2/5. The granulometric composition of ore was determined by the photomapping method.

The amount of coarse fragments was established by indirect measurement of lumps during the drawing and loading of ore. In experiments of the second series, oversize lumps (more than 500 mm) were almost

Table 36

No. of blasts	Drillhole grid, m	Specific explosive consumption in primary breakage, kg/m³	Yield (%) of fractions, mm			
			< 50	50–400	400–500	> 500
		Parallel drill holes				
2	1.2 × 1.2	1.94	25 9	71.5	2.6	—
3	1.4 × 1.4	1.42	20	71.8	8.2	—
5	1.5 × 1.5	1 20	19.7	70 3	10	—
5	1 6 × 1 6	1.09	16	67 3	16 1	0.6
5	1.8 × 1.8	0 86	12.2	61 2	23	3 6
2	2 0 × 2.0	0.7	11 3	51.3	32 4	5
		Fan pattern of drill holes				
3	1.5 × 1.5	1.2	18.5	55	20 4	6.1
3	1.6 × 1.6	1.13	20.3	50.5	22.7	6.5

absent. Such a quality of ore fragmentation is more acceptable technologically as it helps ensure maximum yield of average-size lumps and minimum dilution. Such a quality of fragmentation, conforming to the conditions of autogrinding (without using balls) in crushers, can be obtained by using 65-mm diameter drill holes placed on a grid ranging from 1.5×1.5 up to 1.6×1.6 m.

The relationship between yield of fines, average and coarse lumps and the specific explosive consumption in primary breakage (Fig. 43) was obtained in conformity with the data in Table 26 and Table 36. With an increase in specific explosive consumption for primary breakage from 0.8 to 2 kg/m³ (actual), the amount of fines fraction increased from 12.5 to 25% (twice), the volume of coarse fractions reduced from 27.5 to 2.5% (by 11 times) and the volume of average fraction increased from 57.5 to 70%. The significant increase in average-size lumps was practically assured by an specific explosive consumption from 0.9 kg/m³, while the maximum yield was obtained at 1.2 kg/m³.

Based on the laboratory and in-mine experiments, a marked disparity between the comparative results of blast breakage was observed. For example, at a specific explosive consumption of 1 kg/m³, the error in determination of fine, average and coarse fractions was 15, 36 and 53% respectively (see Fig. 43). Therefore, evaluation of errors during modelling is very essential and must be done by a specially developed methodology.

During experimentation the results of modelling of ore breakage with presplitting of the stoping area [17] were verified; the section consisting of up to 5-m thick steeply dipping ore veins (70–80°), was worked using the new technology of stripping along the strike. The predominant thickness

110

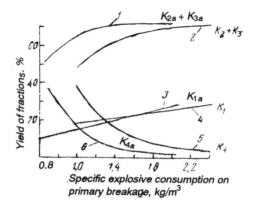

Fig. 43. Dependence of yield of fractions on specific explosive consumption in primary breakage at sizes < 50 (3, 4), 50–400 (1, 2), 400–500 mm (5, 6), 2, 4, 5—model, 1, 3, 6—actual.

of the ore body was 2 m. The enclosing rocks were relatively competent and comprised biotitic sandstones and andesite-diorite prophyrite with a strength of 10–14 and 12–17 on the Protod'yakonov scale respectively.

Horizontal holes of 51 mm diameter were drilled to a depth of up to 15 m into each strip from the drill rise. The strips were worked from the bottom upwards using the stoping complex KOV-25.

The results of breakage of steeply dipping veins of 1.4–2.2 m and 2.8–3.6 m respectively are shown in Tables 37 and 38. In the first case, three drill holes were blasted, in the second case four drill holes, including two cut holes. In this series of experiments, the effect of delay interval between cut holes and perimeter holes on the yield of oversize fragments and ore fines as well as on dilution was established. A comparative evaluation indicated the possibility of relating the results of laboratory modelling with experimental data obtained under field conditions.

As a result of studying the specific features in the working of 1.4–3.6 m thick steep veins by presplitting of the narrow stope area, rational parameters of drilling-blasting operations were determined: charge diameter 47–58 mm, LLR 1.4–1.6 m, delay interval between cut and perimeter charges 12.5–25 ms, specific explosive consumption for primary breakage 1.2–1.4 kg/m^3, which ensured the reduction in yield of ore fines from 18–19 to 10–11%, oversize fragments from 6–7 to 3–4% and breakage of the solid massif from 0.35–0.4 to 0.15–0.16 m. The rational delay intervals for working of steep veins at d_c = 47–58 mm, W_K = 1.5–1.6 m with presplitting are given below:

Vein thickness, m	1.5	2	2.5	3	3.5
τ_{del}, ms	12.5–15	> 15–17.5	> 17.5–20	> 20–22.5	> 22.5–25

Table 37

Volume of broken ore, m³	τ_{del}, ms	Average thickness of vein, m	LLR, m	Average stoping width, m	Distance between cut hole and perimeter holes, m	q, kg/m³	K_{op}, %	K_{dl}, %	H_p, m	R_*, %	No. of experiments
150	15	1.4	1.5	1.63	0.7	2.7	—	21	0.23	1.41	6
170	20	1.4	1.5	1.67	0.7	2.7	—	21.4	0.27	16.2	7
186	15	1.8	1.5	2	0.9	2.1	1.4	17.7	0.20	10	6
205	20	1.8	1.5	2.05	0.9	2.1	1.7	18	0.25	12.2	7
185	15	2.2	1.5	2.37	1.1	1.7	2.1	13.8	0.17	7.2	5
290	20	2.2	1.5	2.40	1.1	1.7	2.6	14.2	0.20	8.3	7

Table 38

Volume of broken ore, m³	T_{del}, ms	Average thickness of vein, m	LLR, m	Average stoping width, m	Distance (m) between drill holes Cut holes and perimeter	Cut holes	q, mg/m³	K_{op}, %	K_{of}, %	H_p, m	R_*, %	No of experiments
265	15	2.8	1.5	3.03	0.9	1	1.8	2.4	16	0.23	7.6	6
140	20	2.8	1.5	2.98	0.9	1	1.8	1.6	15.2	0.18	6	3
176	25	2.8	1.5	3.08	0.9	1	1.8	2	15.8	0.28	9.1	4
155	15	3.2	1.5	3.4	1	1.2	1.57	3	13	0.20	5.9	3
200	20	3.2	1.5	3.36	1	1.2	1.57	2.5	12.5	0.16	4.8	4
300	25	3.2	1.5	3.44	1	1.2	1.57	2.8	13.5	0.24	7.2	6
172	15	3.6	1.5	3.78	1.1	1.4	1.4	4.2	11.5	0.18	4.8	3
280	20	3.6	1.5	3.75	1.1	1.4	1.4	3.6	10.4	0.15	4	7
225	25	3.6	1.5	3.82	1.1	1.4	1.4	4	11.8	0.22	5.7	4

The last series of experiments on ore breakage was conducted in the metal mine named after Lenin, belonging to the Kafansky group. In this mine, the different mineralogical types encountered were pyrite-chalcopyritic and pyrite-chalcopyrite-secondary sulphidic (bornite, chalcosite, covellite) ores. The average ore density was 2.95 g/cm^3, the strength coefficient was 12.1–21.6 on the Protod'yakonov scale, while the predominant values ranged from 12 to 14.

The ore was mined by vertical slices (towards the chamber, compensating cavity and very rarely in a compressed medium) with the help of blasthole charges. Holes were drilled by the rig NKR-100M and were charged with grammonite 79/21 by the charging machine ZMBS-2.

In blocks 1–2 (vein of second apophysis) in the 'Katar' section of level1009, parallel holes up to 32-m deep along the rise side and 20–25 m along the dip side were drilled in on a grid of 2×1.8 m.

In block 4 (zone) in the 'Sever' section of level 927, ore was mined by parallel converging fans of drill holes towards the compensating cavity. There were 5 drill holes of 105-mm diameter in each fan. The depth of ascending holes was 27 m and of descending holes 32 m. The spacing between the fans was 2.7, 2.8 and 3 m and between holes in the fans 0.3–0.6 m, while LLR was equal to 3.5 m.

The adoption of the scheme with parallel converging fans of holes reduced the labour involved in drilling, as five holes in the vertical plane were drilled into the massif by the same rig NKR-100M. Moreover, this drillhole pattern gave rise to concentrations of explosive energy in individual sections of the massif being broken, which in certain cases facilitated adequate mobility of the massif under an increased LLR. However, irregular distribution of explosives within the massif and the absence of directed effect of blast resulted in uneven fragmentation. This in turn was followed by larger yield of oversize fragments, which according to mine data amounted to 50%, relative to the standard lump size of 400 mm.

A drillhole grid of 2×2 m was adopted in block 2 ('Shestaya Severnaya Zheela'—Sixth northern vein) in section 'Sever' at the 927 level. There were four holes in each row; the depth of ascending holes was up to 7 m and of descending holes up to 26 m. Uniform fragmentation was obtained when this scheme was adopted for ore breakage.

As mentioned above, in the case of ore breakage by clustered parallel fans of holes, increase in the yield of oversize lumps is mainly due to the uneven distribution of explosives in the massif. We shall indicate a specific aspect of the phenomenon which could be investigated from the viewpoint of mechanics of brittle fracture.

At a spacing of up to 2.7–3 m between fans of holes, i.e., at a relative distance of 52–58, the rate of enlargement of crack walls is less than 5 m/s and hence the cracks do not branch off. If the LLR is equal to 4 m, the cracks almost do not widen. In such a case, three to five major cracks without branching originate in the ore slice being extracted.

At a spacing of 0.3–0.6 m in-between holes in a fan, i.e., at a relative distance of 6–11, the rate of enlargement of crack walls exceeds 21 m/s. Overcrushing is caused in the near-field zone due to the intensive branching of cracks.

In the 'Sever' section, ore in block 8 (zone) was extracted using parallel holes of 105-mm diameter placed on a grid from 2×2 m to 2.2×2.2 m, with a delay interval between rows of charges of 25 ms. In this block, during 1984, 80,000 T of ore were broken and drawn while the yield of oversize lumps was 12%.

4.2 Establishing a Criterion for Determining Work Done by a Blast

A criterion for work done by a blast was established by Dubnov and colleagues [10], who combined the specific blast energy Q (kJ/kg) and specific volume of gases v_0 (l/kg) to arrive at Qv_0. The consideration of v_0 is essential for setting the experimental data in order. A comparative evaluation of the granulometric compositions of broken ore in the model and in actual conditions revealed, however, that the interrelationship between error in modelling and the criterion Qv_0 is approximated by a broken line. The results of blast fragmentation according to the criterion Qv_0 thereby become unacceptable. The need for developing a new criterion for work available from a blast was self-evident.

An evaluation of results of laboratory modelling and in-mine experiments on ore breakage by comparing the specific explosive consumption, specific blast energy and specific volume of gases released with the quality of fragmentation is a prerequisite to establishing the technoeconomic suitability of different types of explosives. A criterion wherein modelling error is minimalised would undoubtedly be more valid. The results of ore breakage in model blocks (see Table 26) and in actual conditions (see Table 36) served as the basis for such an evaluation. It should be noted that the blasts were conducted with different explosives—in the model blocks by TEN and in the mine by ammonite 6ZhV. It is known that in blasting 1 kg of TEN, 5685 kJ of heat energy is produced and 800 l of gases are released versus 4305 kJ of heat energy and 895 l gases in the case of ammonite 6ZhV. Therefore, the effect of these explosives on the massif would differ. For convenience of comparative evaluation the following notations are introduced: U—specific consumption of explosive energy (kJ/m^3); q—specific explosive consumption (kg/m^3) and V—specific consumption of gases released in the blast (l/m^3). The results from laboratory modelling and in-mine experiments are compared in Tables 39, 40 and 41, according to the yield of fractions of 13.3–17 mm (model) and 400–500 mm (actual) depending on U, q and V. Utilising the experimental data from these tables, correlation tables were generated and using the least square method the coefficients of equations of hyperbolic regression were determined, which are shown in Table 42.

Correlation equations between the yield of fractions K_4 and $U, q. V$ are graphically shown in Fig. 44. The significant deviation between results of ore breakage in the model and in actual conditions is apparent. The error is mainly due to structural features of the massif (presence of discontinuities). Further, variation in error is observed depending on the criterion of work done by a blast selected. It is essential to give a quantitative evaluation of error in modelling. To achieve this, curves showing the relationship of yield of fractions K_4 in the model blocks and actual (Fig. 45) were drawn and a family of straight lines was obtained. The broken line in Fig. 45 corresponds to an ideal case of 'no-error' in the yield of fractions K_4 in the model and the actual. This line originates at the point of intersection of the co-ordinate axes and the tangent of the angle is equal to 1; therefore $K_a = K_m$. As the angular coefficient decreases, the error in modelling increases. Thus the functional relationship $K_a - K_m$ (straight line 1) obtained using the criterion of specific consumption of energy of explosive charge is distinguished by a relatively large error. The straight lines are generally described by an equation of the type $K_a = a + b\, K_m$. Due to the difference in angular coefficient and certain other displacements in the chosen co-ordinate network, different equations for the unknown straight lines are obtained. The graphs of straightline relationships 1 and 2 in Fig. 45 have shifted along the Y-axis by -0.2 (point a_1) and -1 (point a_2) respectively. The constraint equation $K_m - K_a$, which takes into account the criteria of work done by a blast, U, q and V, is given in Table 43.

The values of errors calculated using the graphs (see Fig. 45) and formulae (4.6), (4.7) and (4.8) are given in Table 44. The error can be computed by

$$\Delta k / K_m = (K_m - K_a)/K_m. \tag{4.9}$$

The relationship between $\Delta k/K_m$ and K_m based on the calculated data of Table 44 is shown in Fig. 46. With a decrease in the yield of coarse ore lumps (fractions K_4) from 40 to 50% (due to the increase in specific explosive consumption in primary breakage), the error based on the U criterion (curve 1) increased from 0.77 to 0.95 and based on the q criterion (curve 2) increased from 0.6 to 0.64. Further reduction in the yield of fractions K_4 led to still greater increase in the error of modelling, i.e., the difference between the yield of fractions K_4 in the model and the actual. In the evaluation of results of blast using the criterion of specific consumption of gases V (curve 3), the error is relatively small (0.41). Moreover, it is constant in the area under consideration. The aforesaid

Table 39

Hole diameter, mm	LLR, mm	Spacing between holes, mm	U, kJ/m³	Yield of fractions 13.3–17.0 mm, %	Hole diameter, mm	LLR, m	Spacing between holes, m	U, kJ/m³	Yield of fractions 400–500 mm, %
Model (n = 1:30)					Actual				
1.5	50	50	6,082	27.9	65	2.0	2.0	3,013	32.4
1.7	50	50	7,859	16.0	65	1.8	1.8	3,720	23.0
1.7	50	50	8,356	16.4	65	1.6	1.6	4,652	16.1
1.9	50	50	9,664	10.4	65	1.5	1.5	5,166	10.0
2.0	50	50	10,233	8.5	65	1.4	1.4	6,115	8.2
2.2	50	50	11,879	6	65	1.2	1.2	8,351	2.6
2.4	50	50	13,472	4.4	65	—	—	—	—

Table 40

Hole diameter, mm	LLR, mm	Spacing between holes, mm	q, g/dm³	Yield of fractions 13 3–17.0 mm, %	Hole diameter, mm	LLR, m	Spacing between holes, m	q, kg/m³	Yield of fractions 400–500 mm, %
		Model (n = 1 30)					Actual		
1.5	50	50	1.04	27.9	65	2.0	2.0	0.70	32.4
1.7	50	50	1 40	16.0	65	1.8	1 8	0.86	23.0
1.7	50	50	1 47	16.4	65	1.6	1 6	1.09	16.1
1.9	50	50	1.70	10.4	65	1.5	1 5	1.20	10.0
2.0	50	50	1.80	8.5	65	1.4	1 4	1 42	8.2
2.2	50	50	2.09	6.0	65	1.2	1 2	1.94	2.6
2.4	50	50	2.37	4 4	65	—	—	—	—

Table 41

Model ($n = 1:30$)					Actual				
Hole diameter, mm	LLR, mm	Spacing between holes, mm	V, ml/dm³	Yield of fractions 13.3–17.0 mm, %	Hole diameter, mm	LLR, m	Spacing between holes, m	V, l/m³	Yield of fractions 400–500 mm, %
1.5	50	50	835	27.9	65	2.0	2.0	627	32.4
1.7	50	50	1,092	16.0	65	1.8	1.8	773	23.0
1.7	50	50	1,147	16.4	65	1.6	1.6	976	16.1
1.9	50	50	1,326	10.4	65	1.5	1.5	1,074	10.0
2.0	50	50	1,404	8.5	65	1.4	1.4	1,271	8.2
2.2	50	50	1,630	6.0	65	1.2	1.2	1,736	2.6
2.4	50	50	1,849	4.4					

relationships (see Table 43) undoubtedly indicate that only one of the three equations (4.6), (4.7) or (4.8) can be considered as the key physical relationship. It is evident that equation (4.8) is the key equation since the criterion V differs markedly from U and q [23]. The mechanism of blast breakage of rocks, based on the power factor of the 'piston effect' of detonation products, in compliance with the quasi-static course of the process supports this hypothesis. An evaluation of the work available from a blast using the criterion of specific consumption of gases released showed that this hypothesis corresponds well with the phenomena. Crack formation during a blast is a relatively slow process, dependent on the duration of transmission of energy to the leading portion of the crack for its growth.

The ratio between rate of crack movement in rocks and the velocity of longitudinal elastic wave ranges between 0.029–0.183 in accordance

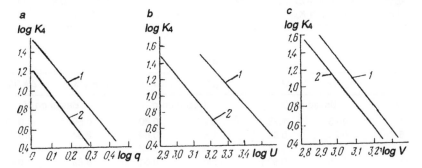

Fig. 44. Dependence of yield of fractions K_4 (1) in the model and (2) in the actual, on (a) specific explosive consumption for primary breakage, (b) blast energy and (c) specific consumption of gases released in the blast in the case of short-delay blasting of 65 mm diameter charges

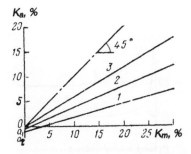

Fig. 45. Relationship between yield of fractions K_4 in the model K_m and in the actual K_a, considering the evaluation of quality of ore fragmentation according to the criteria (1) U, (2) q and (3) V.

Table 42

Evaluation criterion	Object	Equation		
U, kJ/m³	Model	$K_4 = 4.14 \times 10^9 U^{-2.38}$	...	(4.1)
	Actual	$K_4 = 3.07 \times 10^9 U^{-2.54}$	...	(4 2)
q, kg/m³	Model	$K_4 = 37.7 q^{-2.5}$	..	(3 8)
	Actual	$K_4 = 13.18 q^{-2.24}$	...	(4.3)
V, l/m³	Model	$K_4 = 6.061 \times 10^8 V^{-2.49}$	..	(4 4)
	Actual	$K_4 = 5.231 \times 10^8 V^{-2.54}$	..	(4.5)

Table 43

Criterion	Types of equations	Value of coefficient		Equations sought	
		a	b		
U, kJ/m³	$K_{aU} = a_3 + b_3 K_{mU}$	-1	0.25	$K_{aU} = 0.25 K_{mU} - 1$	.. (4.6)
q, kg/m³	$K_{aq} = a_2 + b_2 K_{mq}$	-0.2	0 4	$K_{aq} = 0.4 K_{mq} - 0.2$	... (4 7)
V, l/m³	$K_{aV} = b_1 K_{mV}$	0	0.59	$K_{aV} = 0\,59 K_{mV}$	... (4.8)

Table 44

K_m, %	K_a, %	$\Delta k/K_m$	K_a, %	$\Delta k/K_m$	K_a, %	$\Delta k/K_m$
	Based on U		Based on q		Based on V	
1	−0 75	1.75	0.2	0.80	0 59	0 41
5	0 25	0.95	1.8	0.64	2.94	0.41
10	1.5	0 85	3.8	0 62	5.88	0 41
15	2.75	0 82	5.8	0.61	8.82	0.41
20	4	0.80	7.8	0.61	11.76	0.41
25	5.25	0.78	9 8	0 61	14.70	0 41
30	6.5	0 78	11.8	0.61	17.64	0.41
35	7.75	0.78	13.8	0.61	20.58	0.41
40	9	0.77	15 8	0.60	23.52	0 41

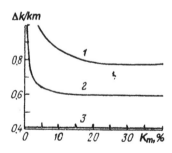

Fig. 46. $\Delta k/K_m$ versus K_m.

with variance in the relative distance between 50 and 5, which makes breakage in the macroscopic volume by the wave effect within the zone of crack formation well-nigh impossible. Hence the products of detonation of the explosive, which affect the duration of 'life' of the rock or ore and the degree of crushing appear to be the most important criterion for determining the work done by a blast.

The criteria U and q cannot be applied while evaluating blast breakage that uses different types of explosives, as the relationship $\dfrac{\Delta k}{K_m} - K_m$ is approximated by a curve. In evaluating the work done by a blast using the criterion V, the error is constant (Fig. 46); this makes it possible to introduce a correction factor j in equation (4.10). In such a case, the quality of breakage in actual practice would be identical to that in the model at,

$$K_a = jK_m. \tag{4.10}$$

For the specific case of modelling of blast breakage, the value of j is equal to 0.41 and depends mainly on the structure (rigidity) of the massif to be broken. The quality of ore or rock fragmentation can be predicted based on ratio (4.10), once we know the coefficient of structural weakening.

It is obvious that evaluation of the work done by a blast using criterion V enables one to compare the results of blast breakage (for example, in the model and in actual practice) and also to compute individually the two component parts of error: the first part, due to the difference in specific volume of gases released by the explosives used, and the second related to variation in the rigidity of the massif. The new criterion, as distinct from the conventional, has a clear-cut physical interpretation; the work done by the piston effect of gases is mainly expended on displacing and breaking the massif.

4.3 Computation of Rational Values for Blast Breakage Parameters

The facts already established and enumerated below serve as the base for computing the rational values for blast breakage parameters in underground metalliferous mining:

—Zone of overcrushing is formed due to the effect of intensive shock in an area confined approximately within two radii of charge;

—In blast breakage, the fractional participation of the blast wave and piston effect of gases for fragmentation is equal to 10–20 and 80–90% respectively;

—In breakage of ores and rocks, the failure front moves from the charge towards the compensating cavity. The rate of crack movement in the massif increases from 25 to 750 m/s with variation of the relative distance from 85 to 4. The rate of enlargement of crack walls (rate of displacement) increases from 10 to 82 m/s with variation in relative distance over 37.6–2.4;

—Ore massif and rocks become fragmented due to the branching of cracks;

—Velocity parameters of breakage of the massif depend mainly on the forces of intermolecular cohesion in ore (rocks) and the degree of manifestation of 'squeezing'.

Data from experimental investigations (see Section 2) show that even at a relative distance of 2.4 the blast wave advances farther than the front of macrocracks. Consequently, the massif breaks (except in the zone of overcrushing) due to the piston effect of gases.

In blast breakage, with an increase in LLR the resistance of ore (rock) for displacement increases proportionately. To maintain the constancy of relative distance while switching over from smaller values of LLR to larger, it is necessary to increase the charge diameter, i.e., to increase the specific consumption of explosive in primary breakage. N.F. Zamesov determined the relationship between increment in specific explosive consumption in primary breakage and the LLR, i.e.,

$$q - q_0 = iW^2, \tag{4.11}$$

where q—specific explosive consumption in primary breakage for the assigned LLR;

q_0—initial specific explosive consumption in primary breakage;

i—proportionality coefficient determined experimentally; its value is assumed depending on the yield of oversize lumps.

The line of least resistance is a function of a series of blast breakage parameters:

$$W = \beta(d_c, \Delta, c_1, c_2, m, K). \tag{4.12}$$

where, d_c —charge diameter, cm;

Δ —loading density, g/cm^3;

c_1 —coefficient based on properties of the massif;

c_2 —coefficient considering the relative strength of explosive;

m —coefficient of closeness of charges;

K —yield of oversize fragments, %.

The coefficient of properties of the massif c_1 is dependent mainly on the extent of discontinuities in it. Determination of c_1 is a complex and difficult operation; hence it is advisable to replace it by another parameter which characterises the condition of the massif fairly accurately. Such a parameter, for example, is the acoustic rigidity of the medium. Knowing the acoustic rigidity of ore and rock massif, the degree of discontinuities can be quantified. Similar work was earlier carried out by Vorotelyak and

Salganik [11] for the conditions of metal mines in Krivoy Rog. They established that $A \times 10^6 = 0.6 - 1.5$ N·f/m^3 for $f = 1.2$-14.3 on the strength scale of Protod'yakonov. Thus, the value of coefficient c_1 is determined taking into account the acoustic rigidity of the medium.

A, N·f/m^3	0.3	0.4	0.5	0.6	0.7	0.8	0.9	1	1.2	1.3	1.4	1.5	1.6
c_1		0.3	0.4	0.5	0.6	0.7	0.8	0.9	1	1.2	1.3	1.4	1.5	1.6

The strength coefficient of explosive c_2 is determined according to computation Table 45. The value of the strength coefficient of ammonite 6ZhV, which under normal conditions provides satisfactory fragmentation of medium-hard ore at a ratio of LLR to the charge radius up to 50, is taken as 1.

The characteristics of grammonite 79/21 are similar to ammonite 6ZhV. By increasing the loading density, however, it is possible to increase the LLR. The value of coefficient c_2 is determined by the ratio of volume of gases released in a blast (l/kg) by the particular explosive and that of ammonite 6ZhV. Certain characteristics of industrial explosives permitted for use in coal and metal mines, and safe with regard to gas and dust explosion [27], are given in Table 45.

Relationship (4.11) shows that the losses of blast energy during breakage of ore (rock) are proportional to the square of distance of charge from the free face. Besides, in order to maintain the rate of displacement of ore along the LLR with an increase in W, it is necessary to increase the charge diameter as well as to ensure the constancy of the ratio,

$$W/r_0 = \text{const}. \tag{4.13}$$

While realising equality (4.13), the rate of enlargement of crack walls and hence the intensity of branching of cracks is maintained. It becomes possible therefore to maintain the quality of ore fragmentation. The LLR and charge radii vary in native and foreign metal mines (Table 46) and their ratio varies over a wide range. In the metal mines of the USSR, the value of ratio (4.13) varies between 43-70 and the average value tends

Table 45

Explosive	Heat produced by the explosion, kJ/kg	Volume of gases released by blast, l/kg
Ammonite 6ZhV	4305	895
Grammonite 79/21	4285	895
Granulite AS-4	4522	907
Granulite AS-4V	4522	907
Granulite AS-8	5191	847
Granulite AS-8V	5233	850
Ammonal M-10	5645	—

124

to 50; this is slightly higher in mines abroad. An analysis of experience gained in blast breakage both at home and abroad and the results of investigations conducted revealed that ratio (4.13) is mainly governed by the properties of the explosive and the massif to be broken and the technique adopted for blast breakage.

The following schema is adopted in selecting rational parameters of ore breakage. It is based firstly on the results of investigations of breakage by blasting in models and their verification under field conditions. Secondly, the kinetic energy of detonation products spent on fragmentation remains constant within the field of rational values of ore breakage parameters and satisfies condition (4.13). Thirdly, depending on the required quality of ore fragmentation, the parameters of blast breakage can be divided into positive factors, if they contribute towards achieving the desired result and into negative factors if they degrade the result. In the present case, drillhole diameter and charge density can be related to the positive factors while the charge grid, i.e., LLR and spacing between charges, is related to the negative.

Based on the aforesaid concepts, the rational value of LLR is determined by the formula

$$W = 50\sqrt{r_0 \Delta c_2 K/(c_1 m)}. \tag{4.14}$$

Table 46

Metal mine	f	W, mm	r_0, mm	W/r_0
Ziryanovskii	9–13	1500	35	43
Komsomolskii	5–8	900	20	45
Gaiskii	9–10	2500–3000	50	50–60
Tekeliskii	6–14	2200–2500	35–40	63
Dzhelambet	14–15	1400	32	44
Leninogorskii	13–14	1500–2400	28–50	48–53
Zodskii	8–12	1500	32	47
Kafanskii	12–14	2200–2500	50	40–50
Cominterns	6–8	2500–3500	42–50	60–70
XX Party Congress	6–8	3000	50	60
Tashtagol	12–16	2500–2800	50	50–60
	12–16	1500–1800	35–40	43–45
*Bautch (USA)	—	1500	37	41
*Climax (USA)	—	1200	37	32
*Hemm (USA)	—	1200	19	63
Brunswick 12 (Canada)	—	1500	20–25	60–75
Jersey (Canada)	—	1200	25	48
*Longsalle (Sweden)	—	1900·	25	76
*Malmberget (Sweden)	—	1500	25	60
*Cobar (Australia)	—	1500	25	60
Mount Isa (Australia)	—	1200	24	50

*Correct spelling could not be verified—General Editor.

Let us determine the LLR for conditions of breaking a massif containing talc-carbonate rocks with a density of $\gamma = 2.4$ g/cm^3 and velocity of propagation of longitudinal wave $C_p = 4100$ m/s. Charges of ammonite 6ZhV of 60-mm diameter are fired with loading density of 0.8 g/cm^3, yield of oversize lumps 4% and coefficient of closeness of charges 1.

The acoustic rigidity

$$A = \gamma C_p = 2.4 \times 410000 = 9.84 \times 10^5 \text{N} \cdot \text{f/m}^3,$$

$$c_1 = \frac{9.84 \times 10^5}{10^6} = 0.984.$$

From Table 45, $c_2 = 1$. LLR, as determined by formula (4.14), is approximately equal to 155 cm.

From formula (4.14) it follows that the charge diameter

$$d_c = \frac{c_1 m W^2}{1250 \Delta c_2 K}. \qquad (4.15)$$

Obviously the charge diameter is determined after taking into account the properties of the explosives and the structure of the ore (rock) massif. In the conditions of underground metalliferous mining, particularly those adopting backfilling, the selection of rational delay interval is of paramount importance in ore breakage.

Short-delay blasting is conducive for concentration of operations enhancing labour productivity and improving the quality of ore fragmentation. It is an effective method of blasting widely adopted in mining enterprises at home and abroad. The existing methods for computing the delay interval are based on the effect of blast wave pulse. An error in computing the delay interval is inherent in the chosen mechanism of blast effect that presupposes blast wave-induced fragmentation of ore. An evaluation of delay intervals conducted by E.G. Baranov revealed that the formulae developed by many authors give significant deviations of value of this parameter from the rational value. Methods of calculating delay intervals based on the rate of failure of the massif are also known. In this case, an attempt is made to quantify the velocity of displacement of failure surfaces (rate of crack movement) as (0.4–0.5) C_p (C_p is the velocity of propagation of longitudinal wave in the massif). Such a relationship only complicates the task. This method of evaluating delay intervals to control the sequence of detonating the charges in time and space is not applicable largely for two reasons: firstly, the real rates of crack movement are significantly low and secondly, their values are not constant due to the increase in dissipation of total energy with relative distance.

The wide usage of short-delay blasting enhances the requirements for evaluation of delay intervals between not only rows of charges, but also charges in a row.

To achieve effective ore fragmentation and to ensure stability of the solid massif, the delay interval between rows of charges can be calculated by the formula

$$\tau_{del} = \tau_{dc} + W/v_T + B/v_b. \qquad (4.16)$$

The first term in this formula represents the duration of phase of dynamic compression of the medium. The second represents the time taken for the movement of crack from the charge to the exposed face with a velocity of 100–640 m/s. In blast breakage, most of these velocities fall within the range 150–300 m/s (see Fig. 16). The third term is the time spent on the formation of a cavity of width B, at an average rate of 10 m/s. The width of cavity between slices is taken to be 0.05–0.8 m, as per the conditions of quality of ore fragmentation and stability of the solid massif. Data from investigations show that the duration of phase of dynamic compression is roughly equal to the duration of movement of major crack from the charge to the exposed face in the slice being broken.

In a multirow short-delay blasting, a specific delay interval is chosen such that the breakage of the subsequent slice is delayed, relative to the preceding one, by an interval adequate for exposing the new face. Here the rate of enlargement of crack walls is an important parameter that determines ore displacement. It is very significant that the velocity of displacement of ore at a constant relative distance does not vary for various radii of charge and LLR, since their ratio usually corresponds to the zone of active failure ($W/r_0 \leqslant 60$). The rational value τ_{del}, determined by formula (4.16) for breaking hard and medium-hard rocks, is equal to 20–30 ms. Therefore, the average delay interval can be taken as 25 ms.

The suggested method of determining τ_{del} in multirow short-delay blasting differs from earlier known methods by the fact that the calculation is based on the dynamics of failure of the real medium. In this case, the degree of complete utilisation of the effect of short-delay blasting is undoubtedly enhanced due to the adequate velocity of ore mobility along LLR, rational redistribution of explosive energy and absence of critical tensile stresses at the sides of the stoping room. In the row-wise long-delay blasting adopted for ore breakage, the blast wave pulse possesses two peak values (maximum and minimum), in which case uneven fragmentation of ore occurs and a large amount of country rocks is broken. In multirow short-delay blasting, the wave pulse attenuates moderately and the tensile stresses at the sides of the room are of low magnitude or, more often, absent. An increase in duration of blast wave affects the mechanism of the accumulatory process, creating conditions for a change in the quality of ore fragmentation.

In modelling blast breakage by the method of equivalent materials, the absence of tensile stresses at the sides of room was indicated when the delay interval was 3.6–3.8 ms. Taking into account ratio (1.20), we obtain for actual conditions $\tau_{del} \approx 25$ ms. This indicates that the rational delay interval for effective ore breakge as well as for ensuring stability of the solid massif happens to be one and the same.

The methodology of evaluation of rational parameters for blast breakage suggested above is related to multirow short-delay blasting of drillhole charges. Working steeply dipping ore veins of average thickness using drillhole charges instead of shothole is advisable due to change in conditions under which explosive charges work. The collar portion of shot holes drilled in the adjacent slice is situated in a zone disturbed by the preceding blast, while the initial state of the massif is maintained at the bottom portion. In ore breakage by drillhole charges, the charges are situated in that part of the massif undisturbed by the blast. Furthermore, the zone of crack formation is formed at the roof of the stope, parallel to these charges. The strength at this part of the massif is relatively low; consequently, it is possible to locate drill holes taking into account the zone of crack formation.

The LLR of a cuthole charge is determined by the ratio (4.13). At a charge diameter of 52 mm, the LLR is equal to 1040 mm (26 × 40), which corresponds to the one adopted in the base mine. Considering the results of ore breakage in actual conditions, the necessity arose to study the influence of LLR of perimeter drillholes W_K on the quality of ore fragmentation in the model. In this series of experiments (see Table 34), the LLR equalled 45–80 mm (in actual conditions 1130–2000 mm) at a constant charge diameter of 2 mm (in actual conditions 52 mm). Data from experimental investigations revealed that in ore breakage with presplitting of a narrow stope area, it is possible to increase the LLR of perimeter drill holes, such that

$$W_K = 1.4\text{--}1.5W_o. \tag{4.17}$$

where W_o is the LLR of cuthole charges determined by (4.13). Violation of these boundary conditions leads to worsening of quality of ore fragmentation. At $W_K \gg 1.5W_o$, a cone of ejection from the cuthole charge and a 'blowout' from the perimeter charges are seen. If $W_K \ll 1.4W_o$, a certain volume of ore is ejected along with the cuthole charge. Experimental data showed that the cuthole charges should be fired during the enlargement process of the walls of the screening cavity, i.e., ore movement.

Firing of the cuthole charges when the width of the cavity is less than 50 mm and more than 80 mm results in uneven ore fragmentation. The suggested method of ore breakage in a narrow stope area involves preblasting of charges placed in perimeter drill holes and subsequent firing of cuthole charges. The effect is different because of the fact that cuthole charges are fired with a delay relative to the perimeter to help reduce the yield of ore fines, to obtain smooth surfaces on the stope

128

walls and to increase labour productivity. This delay is determined by the expression

$$\tau_{del} - t_{dc} + \frac{L}{2v_T} + \frac{B}{v_b}, \qquad (4.18)$$

where L—distance between perimeter holes, m;
v_T—average velocity of crack movement, $v_T = 200$ m/s;
B—width of cavity formed between perimeter holes,
$B = 0.05$ to 0.08 m;
v_b—average rate of enlargement of crack walls; $v_b = 10$ m/s.

The adopted dimension of cavity $B = 0.05$ to 0.08 m allowed adequate movement of particles of the slice being broken when the cuthole charge was fired. Here the effect of blast breakage is associated with the selection of the extent of enlargement of crack walls. In the widening process of walls of the cavity up to 0.05–0.08 m, in the blast of perimeter charges, the slice under breakage moved towards the compensating area, along the LLR. Firing of cuthole charge with a fixed delay interval resulted in augmentation of the kinetic energy of interaction of particles and was accompanied by accelerated movement of ore in the direction of the main compensating area and the cavity as well. As a result of expansion of the contour of massif movement, the zone of branching of cracks expanded, i.e., uniform ore fragmentation was obtained.

Movement of cracks and widening of the walls to the size of a cavity occurred sequentially. Besides, a step-type growth of v_T and v_b was also discernible, albeit there is no interrelation between these parameters. A specific delay interval was determined for each thickness of vein (see Table 40). But the width of the cavity was not the same in the interval L'; it deviated up to 20%. However, it did not influence the results of ore breakage because it is far less than the range of value of widening of walls of the cavity. Hence, the application of (4.17) in the interval L is justified.

Thus, the presplitting of narrow stope area substantially improves the effectiveness of breakage of steeply dipping ore veins of average thickness.

4.4 Technoeconomic Evaluation of Application of Possible Schemes of Breakage by Blasting

The range of rational values of blast breakage parameters that ensures the required quality of ore fragmentation at minimum dilution has been established. The labour inputs and measures under these parameters undoubtedly differ and it is essential to evaluate them when establishing rational values from the point of view of economy. Such an evaluation is more reliable if the parameters of drilling-blasting operations are compared according to the methodology of D.M. Bronnikov. In this, the expenditure on drilling-blasting operations is determined from the expenses for exploiting 1 T of ore.

$$C = C_1 x + (C_2 + C_3)Qx + K_0 C_4 / \lambda_0 +$$
$$+ C_5 / P_m + C_6 + C_7 + V_{dev} C_8 + R_* C_9, \qquad (4.19)$$

where C_1 expenses on drilling 1 m of blast hole, rubles; x—specific length of drilled holes accounting for 1 T of broken ore, m/T; C_2—cost of 1 kg of explosive, including blasting accessories per 1 kg of explosive, rubles; C_3—expenses on transporting and charging 1 kg of explosive, including cost of blasting accessories, rubles; Q—weight of explosive charge per 1 m of drill hole, kg; K_0—yield of oversize fragments from 1-m long drill hole, T; C_4—expenses for crushing 1 T of oversize fragments, rubles; λ_0—yield of ore from 1-m long drill hole; T; C_5—salary (with allowances) of the operator of loading machine or scraper, rubles; P_m—output of machine in loading and hauling ore, T/shift; C_6—amortisation for equipment, rubles/T; C_7—expenses for energy consumed for loading 1 T of ore, rubles; V_{dev}—volume of preparatory workings per 1 T of broken ore, m³; C_8—expenses for drivage of 1 m³ of working, rubles; R_*—dilution of ore, %; C_9—loss due to dilution of ore by 1%, rubles.

According to experimental investigations, the required quality of fragmentation (maximum yield of ore of average fractions) including dilution is obtained when charges placed in drill holes of 65-mm diameter are blasted. The pattern of drill holes might be parallel or the fan type. Thus, the technoeconomic comparison reduces to an evaluation of these schemes of breakage by blasting. The values of initial cost indices of stoping, obtained by processing the records of the production section and observations during drilling, discharge of ore and other operations are given below:

Expenses for drilling 1 m of hole (C_1), rubles ... 2.35
Cost of 1 kg explosive and blasting accessories (C_2), rubles ... 0.65
Expenses for transporting and charging 1 kg of explosive
and blasting accessories (C_3), rubles ... 0.1
Expenses for crushing 1 T of oversize lumps, rubles ... 0.16
Expenses for loading and hauling 1 T of ore $(C_5 + C_6 + C_7)$,
rubles ... 0.34
Expenses for drivage of 1 m³ of working (C_8), rubles:
with parallel drill holes ... 20.05
with fan-type drill holes ... 26.39
Loss due to dilution of ore by 1% (C_9), rubles ... 0.07

The main technoeconomic indices of ore breakage by vertical slices using charges placed in drill holes of 65-mm diameter in the trial block 2/5 are given below.

Drillhole pattern	Parallel	Fan type
LLR, m	1.6	1.6
Spacing between drill holes, m	1.6	1.6
Specific length of drill holes, m/T	0.16	0.23
Yield of ore per 1 m drill hole, T	6	4.5
Specific explosive consumption in primary breakage, kg/T	0.43	0.45
Specific explosive consumption in secondary blasting, kg/T	0.004	0.035
Yield of oversize lumps, %	0.6	6.5
Yield of oversize lumps per 1 m of drill hole, T	0.03	0.27
Weight of explosive charge contained in 1 m of drill hole, kg	2	2
Specific volume of preparatory workings, m^3/T	0.042	0.024
Dilution, %	3	5

Depending on the drillhole pattern, these indices might differ. The main disadvantages of fan-type drill holes are: variable distance between the ends of drill holes (due to deviation), leading to uneven ore fragmentation; incomplete utilisation of drill holes (due to undercharging); breaking into solid massif (dilution) due to deviation of drill holes and increased expenditure due to additional volume of drilling operations. These disadvantages can be overcome by using the parallel hole pattern. In this case, the drilling and discharge horizons can be combined, which reduces the volume of preparatory workings and drilling of holes. Nevertheless, the fan-type drillhole pattern is still widely used. This is probably because frequent shifting of the drill machines is necessary in the parallel pattern of drilling. Such a situation arises from the lack of conformity between constructive design of method of working and the chosen parameters of blast breakage. For example, in many underground metal mines of the Krivoy Rog basin in which wide ore bodies are worked, it is possible to adopt block caving instead of sublevel stoping. In consequence, the holes could be drilled from a single drilling horizon and operations related to the shifting of drill machines could be minimalised or the field of application of free-steered equipment widened; furthermore, the volume of drifting operations would reduce due to the reduced number of sublevel workings.

Based on experimental results (see Table 36), and applying equation (4.19), the economic viability of blast breakage was established for both

Table 47

Operation	Cost (rubles/T) of breakage by blasting with drillhole charges of 65-mm diameter	
	Parallel pattern	Fan pattern
Drilling	0.38	0 54
Blasting	0.24	0.35
Breaking oversize lumps	0 01	0.01
Delivery [of explosives etc., to site]	0.34	0.34
Drifting	0.84	0.63
Loss due to dilution	0.21	0.35

parallel and fan-type drill holes. The cost of ore breakage in the fan pattern is 10% higher compared to the parallel hole scheme (Table 47).

In the course of investigations on blast breakage while working block 1/98, it was seen that the ore was prone to consolidation and uneven fragmentation. This led to rejection of the shrinkage method of working using 105-mm diameter drill holes and a field trial was conducted with the room-and-pillar method from sublevel drifts and subsequent backfilling. Satisfactory results were obtained in block 2/5 where charges placed in 65-mm diameter holes were used. In order to bring about major refinements in the technology of mining ore deposits, certain compositions of material for filling the stoped-out area were developed that facilitate simultaneous extraction of ore from several levels. This in turn contributed to expanding the design capacity of the mine and improving the techno-economic indices.

The refinements in breakage by blasting helped lower expenses for extraction of ore. If expenses for drilling-blasting operations in the case of breaking ore by charges of 105 and 41-mm in diameter were 2.58 and 1.47 rubles/T respectively, then by switching over to the new method of ore breakage by charges of 65 and 41 mm in diameter, the expenses decreased by 0.86 and 0.24 rubles/T. Furthermore, the dilution decreased from 11.2 to 5.6%.

In the central part of the deposit, where ore and country rocks are unstable, the ore is mined by using 41-mm diameter charges.

Ore is broken using charges of diameter 65 mm (40%) and 41 mm (60%) at the stage of fully commissioning the design capacity of the base mine. The annual economic contribution due to implementation of the results of these investigations has been estimated to be 955,200 rubles.

4.5 Breakage by Blasting and Perfecting Technology of Ore Extraction

In the technical reconstruction of underground metalliferous mines, considerable attention is being paid today to perfecting the technology of mining vein-type deposits of rich and average-value ores, which are extracted under complex geomining conditions using less productive technology. In recent years, in working steeply dipping ore veins higher technoeconomic indices have been achieved by adopting the following methods of working: sublevel drifts; horizontal slices with backfill [26]; stripping by machine complexes mounted on monorails [17] etc. Such methods notwithstanding, the expenses on ore breakage, its delivery and supporting the stoped-out area amount to 75–90% of the total cost of ore extraction.

Productivity and concentration of extraction processes depend mainly on the parameters of breakage by blasting and blasting patterns. However, neither physical representations nor accumulated experience provide the engineers with methods for augmenting productivity of blast breakage and consequently that of the technology of ore extraction per se. When the stoped-out area was filled with cemented fill material, though resulting in upgraded rate of recovery, it did not lead to a significant reduction in dilution; contrarily, dilution often tended to increase. For example, with the adoption of drill holes of smaller diameter instead of shot holes in working steeply dipping ore veins of average thickness, the dilution increased sharply due to blast-induced mechanical effects at the back of the designed contour. Research is being directed towards bridging this lacuna as well.

Mining practice in metal mines confirms the need for enlarging the scope of investigations into blast-induced mechanical effects so as to bring about further refinements in the technology of ore extraction.

Zodskii metal mine extracts the central part of the Zodskii deposit, consisting of hydrothermally altered and brecciated ore zones related to tectonic features. The presence of various apophyses and narrow veins, often merging, are characteristic of these zones. The metal values are unevenly distributed in the ore. Due to the absence of clear-cut geological contacts, the boundary of economic mineralisation is established by sampling. The ore bodies are distinguished by complex morphology and their thickness is in the range 1–25 m and the angle of dip commonly in the range of 70–90°. Country rocks are largely represented by gabbro, serpentinised peridotites and sparsely distributed hardwood rocks. In the proximity of ore bodies, at a distance up to 5–10 m, the country rocks were subjected to intensive hydrothermal alterations, which resulted in loss of strength properties. The strength coefficient of the country rocks and ores on the Protod'yakonov scale is 3–20, with values of 8–12 predominating. The ores are prone to consolidation. The stability of ore and country rocks is uneven dependent on the extent of discontinuities in the massif. Adjacent sections might contain both competent and incompetent ores and rocks.

Under such conditions, the need arose for selecting a method of working compatible with improved quality, comprehensive recovery of precious ore and safety of operations. Under the given geomining features of the deposit, the following methods of working are likely candidates: (1) system of working with descending slices and filling—in sections with incompetent ores and rocks; (2) room-and-pillar method of working with extraction of ore from sublevel drifts and subsequent backfilling—in sections with competent ores and rocks. These modifications, if adopted, would enable establishment of the contours of the ore bodies by sampling from the slice or sublevel drifts, as well as rational utilisation of the parameters of blast breakage and blasting patterns.

However, lacunae still exist in evaluating the properties of fill mixtures, which limit the scope of utilisation of blast breakage parameters and, in turn, the technology of ore extraction with backfilling. For example, a fundamental concept such as the standard strength of a fill structure needs to be refined.

While considering the variation of dynamic stress on the static strength function of material, a practically significant solution to the problem of strain-strength properties of a solid massif (see Section 2) was obtained. The change in dynamic stress in a solid (filled) massif during breakage of ore by blasting is caused by failures in microscopic volumes, which are irreversible in nature. Here the wall rocks impede the swelling of the medium from the moment stress σ_T is attained, which corresponds to the boundary of formation zone of irreversible microcracks, resulting in increased strength of the solid massif (or structure). The value of σ_t, characterising the boundary of discontinuities in the medium, is a vital material parameter. The ratio σ_T/R_{st} is proportionate to the variation in material strength. A new mixture for filling the stoped-out area in metal mines, comprising volcanic porous rocks and in particular lithographic pumice stone of high compressive strength properties, was developed [2]. The properties of capillary-porous lithographic pumice, such as enhanced activity and transportability, facilitated the introduction of technology of filling operations in the exploitation of deposits under complex geomining conditions. Cementing mixtures based on lithographic pumice were introduced for the first time on a large scale in underground metal mining practice while working the Zodskii deposit. This facilitated continuous unhindered extraction at various levels by descending slices with backfill.

Field investigations of artificial massifs containing lithographic pumice enabled the following observations. The strength indices of such fill structures are 2.5–3 times higher than the initial strength of a month-old massif.

The recommended compositions of cementing mixture pose an insurmountable obstacle for dynamic cracks, especially during breakage by blasting. Under such conditions, the normative strength of a fill structure does not exceed 6.7 MPa. The residual strength of recommended fill mixtures is 40–50% of the ultimate strength, which is distinctly different from the post-failure characteristic of a rock massif.

Two circumstances helped to bring about radical improvement in the technology of mining the Zodskii deposit. Firstly, the use of highly effective fill mixtures and secondly, introduction of free-steered equipment [22].

The design assignments carried out by VNIPI Gortsevtmet and Armniprotsvetmet for the Zodskii metal mine are based on the recommendations of IPKON AN USSR and Armniprotsvetmet regarding adoption of methods of working with backfilling a stoped-out area. Field trials with these methods of working were begun in 1969 and were completed in 1977, which enabled timely attainment of the design capacity of the mine. The major part of the ore is exploited by the upgraded improved method of working with descending slices and backfill with deployment of free-steered equipment.

Depending on the thickness of the ore body, two technological schemes of extraction have been adopted: stripping along the strike under the cover of the artificial massif made of cemented fill for an ore body up to 12–15 m in thickness. Still thicker bodies are worked by stripping across the strike. The span of exposure of the artificial massif has been increased from 6 to 6–8 m. The height of the stoping blocks is 45–55 m, length 170–200 m and width equal to the thickness of the ore body. The block is prepared mainly by driving a transport incline (inclination 8–10°), ore passes, access drift to the slice, haulage drive, ore drawal horizon, rock fill and air raise. The location of the transport incline was determined by the stability of the rocks in the footwall or hanging wall.

Ore is exploited by shothole charges of 41-mm diameter and 3-m deep. This provides for adequate fragmentation of the massif without yielding oversize lumps.

Stoping at all stages is carried out using compressed-air-driven, free-steered complexes, comprising loading-hauling machine, drill rig and a utility vehicle. The technoeconomic indices of the method of working with descending slices and deployment of free-steered equipment are given below:

Specific volume of preparatory-development operations, m^3/1000 T	...	23–30
Output per manshift in this method of working (including fill), T/shift	...	20–35
Output of load-haul-dumper Toro 200D, T/shift	...	150–200

Specific explosive consumption in primary breakage, kg/m^3	...	0.9–1
Consumption of timber (for supports), m^3/m^3	...	0.03
Loss, %	...	1.5
Dilution, %	...	6–10

Notwithstanding the high technoeconomic indices, significant potential exists for improving the output per manshift. Massive reserves of active lithographic pumice of the Tsovaksk deposit situated at a distance of 35 km from the mine and the possibility of utilising PEF from the Araratsk cement-asbestos plant facilitate timely fill operations and improve the work organisation for exploiting these reserves.

It should be noted that the main effect of perfected technology of ore exploitation with backfill in the descending order is associated with the use of lithographic pumice having high hydraulic (chemical) activity (80–84.5 mg CaO/g sand). The aforesaid activity was determined by the methodology of V.R. Israelyan. His investigations established that the contact zone of portland cement binder (including PEF) and pumice is a finely dispersed, reactive border/edging that differs from the matrix-type cement brick in elastoplastic properties and serves the role of retarder of initiated and developed shrunk cracks. Owing to the high viscosity of the magmatic melt of acidic composition and conditions of its outflow to the earth's surface, the glassy-phase pumice possesses a fibrous structure which enhances the deformation properties of the material. Thus a fill material based on lithographic pumice, despite the high glass content (80–85%), does not fail in a brittle manner.

In sections with competent ore and rock, the ore is worked from sublevel drifts by drillhole charges of 55–65-mm diameter and placed on a grid of 1.5 × 1.5–1.6 × 1.6 m with a delay interval of 25 ms between the rows of charges. This helps to increase the yield of average fractions up to 70%, while damage to the solid massif remains minimal. The delay interval is determined by formula (4.17).

Shaumyansk metal mine. The deposit consists of ore-bearing veins containing polymetals, quartz and carbonate dipping at an angle of 65–85°. The thickness of the ore bodies is 1.16 m with a thickness between 1.3–1.5 m predominant. The strength coefficient of the ore is 10–15 on the Protod'yakonov scale and density 2.95 g/cm^3. The velocity of longitudinal wave in the ore is 3000–4450 m/s. Country rocks are represented by coarse- and medium-blocky andesitic, dacitic, quartzitic porphyrites. The strength coefficient of the rocks is 14–17. Both the ore and wall rocks are by and large competent. Sections with developed passive tectonics contain injected breccia of quartzitic, andesitic dacites of average strength.

In the absence of precedent experience, the method of working the Shaumyansk deposit was explored through field trials. Here it was necessary to consider the following major peculiarities which limit the field of application of the method described above. Firstly, the ore bodies are situated close to each other, their elements of deposition are not constant and the metal values are unevenly distributed in the ore. Apophyses and veins, swelling and thinning considerably complicate the technology of extraction. Combined working of ore bodies is not advisable. Secondly, it is necessary to protect buildings, structures and fertile fields on the surface. The technology of ore extraction with cemented backfilling of the stoped-out area fulfils these conditions to some extent. It is possible, using this technology, to improve the quality and totality of recovery and also to solve a series of problems associated with the negative impacts of mining activities on the surrounding environment. On the other hand, the need for early commissioning of the design capacity in the mine calls for speeding up of mining operations, which could be achieved by mechanisation of principal technological processes of extraction with breakage of ore by drillhole charges of small diameter. An evaluation conducted under the geomining and geotechnical factors revealed that given the conditions of the Shaumyansk metal mine, the room-and-pillar method of working with breakage of ore from sublevel drifts and subsequent backfilling with a cementless hardening mixture would be the most appropriate.

The adoption of highly effective mechanised technology of working a vein-type deposit poses yet another important problem associated with the transition to drill holes of smaller diameter in place of shot holes for ore breakage. An increment in diameter and length of charge weights combined with a sharply growing effect of 'wedging' at the narrow stope face, would alter the nature and conditions of the charge effect on both the ore massif and the host (fill structure) beyond the design contours of the stoping area. The requirements vis-à-vis results of blast breakage remain the same, however: required quality of fragmentation causing minimum damage to the enclosing rocks (fill structure). Hence, a method of controlling the blast effects on ore breakage in a narrow stoping area with preblasting of charges in the perimeter holes and subsequent blasting of cut holes was developed and introduced. In this method the charges in cut holes are fired at a rational delay interval relative to the perimeter charges as established by formula (4.18) and the perimeter holes are pulled forward by $(0.4-0.5)\ W_0$ compared to the cut holes, for the purpose of reducing the yield of ore fines, obtaining relatively smooth-surfaced walls, improving output per manshift and reducing explosive consumption. A cavity between the volume of ore being broken and the massif forms as a result of preblasting of perimeter charges. The rational values of blast breakage parameters are as follows: diameter of charges 47–58 mm, LLR

1.4–1.6 m, delay interval between cut and contour charges 12.5–25 ms with variation in thickness of ore body from 1.5 to 3.5 m, which provides, as compared to subsequent contouring, reduced yield of ore fines, over-size lumps and reduced effect of blast on the solid massif. Formation of a cavity of required size and lag in initiation of charges compared to contour charges increases the kinetic energy of interaction of particles of the mas-sif being broken, due to the accelerated movement of ore in the directions of both the compensating area and the cavity. This method was tested in one of the metal mines of Uzbek SSR as part of experimental field trials of a new technology for mining vein-type deposits by strips along the strike and deployment of monorail-mounted machine complexes [30]. The tech-noeconomic indices of ore breakage in steeply dipping veins of average thickness by drillhole charges of 52 mm diameter are given below:

Thickness of extraction, m	...	2.8–3.2
Depth of drill holes, m	...	10.5
Specific metres drilled, m/m^3	...	0.6
Specific explosive consumption, kg/m^3	...	1.6–1.8
Output per manshift of a face worker, m^3/shift	...	26.2
Granulometric composition of ore, %:	...	
yield of oversize lumps (> 300 mm)	...	2.1
yield of average fractions (50–300 mm)	...	83.7
yield of small lumps (< 50 mm)	...	14.2

As a result of adopting this efficient method of ore breakage, the specific metres drilled was reduced by more than 30% and the output per manshift of face worker rose by 16%.

However, these results were obtained in breakage of isolated slices. Many more experiments need to be carried out in the Shaumyansk metal mine to effect a transition to breakage by 3–4 rows of drillhole charges of small diameter for working steeply dipping veins of average thickness.

Row-wise working by drillhole charges with short delays is recommended in relatively thicker sections of the ore bodies in the Shaumyansk metal mine. The main parameters of blast breakage, computed by formulae (4.14), (4.15) and (4.16), are as follows: diameter of charges 52–60 mm, placed in a grid of 1.2 × 1.2 m–1.4 × 1.4 m, with delay interval between rows of charges 25 ms.

The thrust of investigations and the practical utilisation of results do not exhaust all the possibilities of the methods suggested in this book. With advances in blast breakage, undoubtedly new tasks will arise whose solution could emerge through application of the methodology of experi-mental investigations proposed herein.

References

1. A.S. SSSR No. 815292 MKI E21 C37/00. Sposob otboiki poleznykh iskopaemykh (Method of breaking minerals). E.G. Baranov, V.N. Mosinets, E.M. Podoinitsin et al. Bulletin no. 11, 23.03.81.

2. A.S. SSSR No. 389271 MKI E21 F15/00. Sostav dlya zakladki vyrabotannogo prostranstva rudnikov (Composition of material for filling the stoped-out area in metal mines). M.I. Petrosyan, A.A. Babayan, V.G. Panosyan et al. Bulletin no. 29, 05.07.73.

3. A.S. SSSR No. 670859 MKI G01 No. 21/16. Sposob registratsii rosta treshchin pri vzryvnom razrushenii gornykh porod (Method of recording crack growth during breakage of rocks by blasting). M.I. Petrosyan and T.G. Gasparyan. Bulletin no. 24, 30.06.79.

4. K.K. Arbiev. 1985. Sostoyanie i razvitie rudnoi bazy tsvetnoi metallurgii SSSR (The status and development of mineral base of non-ferrous metallurgy in the USSR). *Gornyi Zhurnal*, no. 3, pp. 3–7.

5. E.G. Baranov. 1982. Puti intensifikatsii protsessov otboiki, drobleniya i izmel'cheniya zhelezhnykh rud (Means for intensifying the procésses of breakage, crushing and milling of ferrous ores). *Gornyi Zhurnal*, no. 8, pp. 40–42.

6. N.M. Belyaev. 1976. Soprotivlenie materialov (Strength of Materials). Nauka, Moscow.

7. V.A. Borovikov and I.F. Vanyagin. 1976. K raschetu parametrov volny napryazheniya pri vzryve udlinennogo zaryada v gornykh porodakh (Computation of parameters of stress wave in a blast of elongated charge in rocks). In: *Vzryvnoe Delo*, 76/33, pp. 74–82. Nedra, Moscow.

8. Vliyanie vzryvnogo nagruzheniya na fizicheskie i tekhnologicheskie kharakteristiki zhelezhistykh kvartsitov (Impact of blast loading on the physical and technological characteristics of ferruginous quartzites). N.Ya. Repin, V.A. Zraichenko, A.I. Potapov et al. *Izv. Vuzov. Gornyi Zhurnal*, 1984, no. 2, pp. 47–51.

9. A.I. Voronin, V.G. Moiseev and V.S. Savelkov. 1988. Opyt vedeniya zakladochnykh rabot na osnove otkhodov proizvodstva (Backfilling practice utilising industrial wastes). *Gornyi Zhurnal*, no. 6, pp. 41–43.

10. L.V. Dubnov, N.S. Bakharevich and A.I. Romanov. 1988.

Promyshlennye vzryvchatye veshchestva (Industrial Explosives). Nedra, Moscow.

11. Instruktivno-metodicheskie ukazaniya po vyboru ratsional'nykh parametrov pri podzemnoi ochistnoi vyemke na shakhtakh krivorozhskogo basseina i ZZhRK-1 (Systematic instructions for selecting rational parameters of underground stoping in mines of the Krivoi Rog basin and ZZhRK-1). G.A. Vorotelyak, V.A. Salganik, N.P. Oleinik et al. NIGRI, Krivoi Rog, 1977.

12. A.L. Isakov and E.N. Sher. 1983. Zadacha o dinamike razvitiya napravlennykh treshchin pri shpurovom vzryvanii (Problem of dynamics of development of oriented cracks in shothole blasting). *Fiziko-tekhnicheskie Problemy Razrabotki Poleznykh Iskopaemykh*, no. 3, pp. 28–36.

13. N.V. Mel'nikov. 1978. Problem kompleksnogo ispol'zovaniya mineral'nogo Sir'ya (Problems of complex utilisation of raw minerals). In: Gornaya Nauka i Ratsional'noe Ispol'zovanie Mineral'no-syr'evykh Resursov, pp. 14–28. Nauka, Moscow.

14. V.N. Nikolaevskii. 1984. Mekhanika poristykh i treshchinovatykh sred (Mechanics of Porous and Fissured Media). Nedra, Moscow.

15. V.S. Nikiforovskii and E.I. Shemyakin. 1979. Dinamicheskoe razrushenie tverdykh tel (Dynamic Failure of Solid Bodies). Nauka, Novosibirsk.

16. V.S. Nikiforovskii. 1976. O kineticheskom kharaktere khrupkogo razrusheniya tverdykh tel (Kinetic nature of brittle failure of solid bodies). *Zhurnal Prikladnoi Mekhaniki i Tekhnicheskoi Fiziki*, no. 5, pp. 150–157.

17. Novaya tekhnologiya razrabotki zhil'nykh mestorozhdenii i metodicheskie ukazaniya po ee primeneniyu (New technology for exploitation of vein type deposits and systematic directions for its application). IPKON AN SSSR, Moscow, 1981.

18. I.I. Noskin, G.M. Baturina and V.F. Ponomarev. 1987. Vnedrenie niskhodyashchei sloevoi vyemki rudy na Zmeinogorskom rudnike (Introducing the method of extracting ore by descending slices in the Zmeinogorsk metal mine). *Gornyi Zhurnal*, no. 5, pp. 33–35.

19. A prognozirovanii razrusheniya gornykh porod (Predicting breakage of rocks). S.N. Zhurkov, V.S. Kuksenko, V.A. Petrov et al. *Izvestiya AN SSSR. Fizika Zemly*, 1977, no. 6, pp. 11–18.

20. A.A. Pashchenko, B.M. Vypolzov and S.Kh. Tulyaev. 1989. Bestsementnye zakladochnye smesi na osnove promyshlennykh otkhodov (Cementless fill mixtures from industrial wastes). *Gornyi Zhurnal*, no. 4, pp. 23–25.

21. M.I. Petrosyan. 1979. Ustanovlenie skorosti rosta treshchiny v gornykh porodakh (Determination of growth rate of cracks in rocks). *Promyshlennost' Armenii*, no. 8, pp. 48–51.

140

22. M.I. Petrosyan, N.F. Zamesov and S.S. Arzuymanyan. 1983. Sovershenstvovanie sistemy razrabotki gorizontal'nymi sloyami s zakladkoi (Perfecting the method of working with horizontal slices and backfill). *Promyshlennost' Armenii*, no. 7, pp. 31–35.

23. M.I. Petrosyan. 1980. Ustanovlenie fizicheskogo kriteriya otsenki raboty vzryva pri otboike rudy (Determination of physical criterion for evaluating the work done by a blast in breaking ores). *Promyshlennost' Armenii*, no. 5, pp. 41–44.

24. M.I. Petrosyan. 1985. K izucheniyu deformatsionno-prochnost-nykh svoistv napryazhyenno-deformiruemogo massiva pri vzryvnoi nagruzke (On the study of strength-deformation characteristics of stressed-deformable massif under blast loading). Tezisy Dokladov VII Vsesoyuznogo Soveshchaniya 'Fizicheskie Svoistva Gornykh Porod pri Vysokikh Davleniyakh i Temperaturakh'. Erevan, pp. 113–114.

25. M.I. Petrosyan. 1985. O dolevom uchastii vzryvnoi volny i porshnevogo deistviya gazov v razrushenii napryazhenno-deformiruemogo massiva (Fractional participation of blast wave and piston effect of gases in breaking a stressed-deformable massif). *Fiziko-tekhnicheskie Problemy Razrabotki Poleznykh Iskopaemykh*, no. 5, pp. 61–68.

26. M.I. Petrosyan and V.A. Avetikyan. 1984. Rezul'taty issledovaniya deformatsionnykh svoistv smesei dlya zakladki vyrabotannogo prostranstva rudnikov (Results of investigations into the deformation properties of compositions used for filling stoped-out area in metal mines). In: Sovershenstvovanie Tekhnologii Gornykh Rabot pri Kompleksnom Ispol'zovanii rud Tsvetnykh Metallov, pp. 86–95. Erevan.

27. Z.G. Pozdnyakov and B.D. Rossi. 1977. Spravochnik po promyshlennym vzryvchatym veshchestvam i sredstvam vzryvaniya (Handbook on Industrial Explosives and Blasting Accessories). Nedra, Moscow.

28. Razrushenie i deformirovanie tverdoi sredy vzryvom (Breakage and deformation of a hard medium by blasting). *Vzryvnoe Delo* No. 76/33. V.N. Radionov (ed.). Nedra, Moscow, 1976, p. 6.

29. Svoistva gornykh porod pri raznykh vidakh i rezhimakh nagruzheniya (Properties of rocks subjected to different types and regimes of loading). A.I. Beron, E.S. Vatolin, M.I. Koifman et al. Nedra, Moscow, 1984.

30. Sovershenstvovanie otboiki rudy skvazhinami v usloviyakh zhil'nykh mestorozhdenii (Perfecting ore breakage by drill holes under conditions of vein type deposits). Yu.P. Galchenko, S.S. Gasparyan, V.M. Zakalinskii et al. *Fiziko-tekhnicheskie Problemy Razrabotki Poleznykh Iskopaemykh*, 1986, no. 5, pp. 63–68.

31. Sovershenstvovanie zakladochnykh rabot na Nikolaevskom rudnike (Perfecting backfilling operations in the Nikolaevsk metal mine).

K.N. Svetlakov, S.A. Atmanskikh, A.T. Mogil'nyi et al. *Gornyi Zhurnal,* 1989, no. 6, pp. 44–45.

32. S.P. Timoshenko and J. Goodyear. 1975. Teoriya uprugosti (The Theory of Elasticity). Nauka, Moscow.

33. A.M. Turichin. 1976. Elektricheskie izmereniya neelektricheskikh velichin (Electrical Measurements of Non-electric Quantities). Energiya, Leningrad.

34. Yu.I. Fadeenko. 1977. Vremennye kriterii razrusheniya vzryvom (Time-related criteria of breakage by blasting). *Zhurnal Prikladnoi Mekhaniki i Tekhnicheskoi Fiziki,* no. 6, pp. 154–159.

35. E.I. Shemyakin. 1978. K izucheniyu mekhanizma razrusheniya prochnykh gornykh porod udarnymi nagruzkami (On studying the mechanism of breakage of hard rocks by impact loads). In: Voprosy mekhanizma razrusheniya gornykh porod. IGD SO AN SSSR, Novosibirsk, pp. 3–14.

36. E.N. Sher. 1982. Primer rascheta dvizheniya radial'nykh treshchin, obrazuyushchikhsya pri vzryve v khrupkoi srede v kvazistatich-eskom priblizhenii (Illustration of computing the movement of radial cracks formed due to a blast in a brittle medium with a quasistatic approxima-tion). *Fiziko-tekhnicheskie Problemy Razrabotki Poleznykh Iskopaemykh,* no. 2, pp. 40–42.

37. E.N. Sher. 1982. Dinamika khrupkikh treshchin s prilozheniyami k mekhanike vzryva (Dynamics of brittle cracks as applied to blasting mechanics). Avtoref. Dokt. Diss. Novosibirsk.

38. Effektivnost' niskhodyashchei sloevoi vyemki s tverdeyushchei zakladkoi (Effectiveness of extraction by descending slices with cemented backfill). G.M. Pustokhin, A.I. Raish, K.N. Svetlakov et al. *Gornyi Zhurnal,* 1989, no. 7, pp. 24–26.

39. D.R. Curran, D.A. Shockey, L. Seaman and M. Autin (Stanford Research Institute, Menlo Park, California). Mechanisms and models of cratering in earth media. In: Impact and Explosion Cratering. D.J. Roddy, R.O. Repin and R.B. Merrill (eds.). Pergamon Press, New York, 1977, pp. 1057–1087.

*For Product Safety Concerns and Information please contact
our EU representative GPSR@taylorandfrancis.com Taylor & Francis
Verlag GmbH, Kaufingerstraße 24, 80331 München, Germany*

T - #0041 - 160425 - C0 - 229/152/9 [11] - CB - 9789061919025 - Gloss Lamination